T0353473

From Pinch Methodology to Pinch–Exergy Integration of Flexible Systems

Thermodynamics – Energy, Environment, Economy Set

coordinated by
Michel Feidt

From Pinch Methodology to Pinch–Exergy Integration of Flexible Systems

Assaad Zoughaib

First published 2017 in Great Britain and the United States by ISTE Press Ltd and Elsevier Ltd

ISTE Press Ltd
27-37 St George's Road
London SW19 4EU
UK

www.iste.co.uk

Elsevier Ltd
The Boulevard, Langford Lane
Kidlington, Oxford, OX5 1GB
UK

www.elsevier.com

Notices

Knowledge and best practice in this field are constantly changing. As new research and experience broaden our understanding, changes in research methods, professional practices, or medical treatment may become necessary.

Practitioners and researchers must always rely on their own experience and knowledge in evaluating and using any information, methods, compounds, or experiments described herein. In using such information or methods they should be mindful of their own safety and the safety of others, including parties for whom they have a professional responsibility.

To the fullest extent of the law, neither the Publisher nor the authors, contributors, or editors, assume any liability for any injury and/or damage to persons or property as a matter of products liability, negligence or otherwise, or from any use or operation of any methods, products, instructions, or ideas contained in the material herein.

For information on all our publications visit our website at http://store.elsevier.com/

British Library Cataloguing-in-Publication Data
A CIP record for this book is available from the British Library
Library of Congress Cataloging in Publication Data
A catalog record for this book is available from the Library of Congress
ISBN 978-1-78548-194-9

Printed and bound in the UK and US

Contents

Foreword

This book proposes a novel vision of thermodynamics as applied to the engineering of systems and processes subject to energy and mass transfer and conversion.

The methodology presented in this book aids in the design of systems and processes industry needs with improved sustainability and efficiency while considering the practical and economical feasibility.

This book is, therefore, at the heart of the *Thermodynamics – Energy, Environment, Economy* Set.

It answers a need for students at the bachelor level but also for engineers and researchers.

This book presents the main research work lead by the author within the Department of Energy and Processes (DEP) at Mines ParisTech. I had the chance to follow and participate in part of these works, always while working closely with the industry.

The author already teaches about these new methodologies in many of his classes in different Master programs which testimony the actual character of these works and their industrial applicability.

Without a doubt, this work will be followed by many developments and perspectives, the possibilities of which are discussed in the Conclusion of this book.

I sincerely hope that this work finds the success it deserves!

Michael FEIDT
Emeritus Professor
University of Lorraine, France

Introduction, Context and Motivations

In this chapter, contextual elements are presented to show the link between world systemic constraints (demographic and economic growths), resource consumption (focusing mainly on energy) and environmental issues (climate change and upcoming resources scarcity) related to current production and consumption patterns.

This leads to an anticipated situation that is not sustainable unless the world changes its paradigms in terms of energy supply and consumption. Energy efficiency is one of the cornerstones of energy systems and process design for the future, while their flexibility is the other, considering the future picture of the energy supply system. Indeed, the introduction of renewable and renewed energy will introduce a higher level of design and management complexities with a higher need for process flexibility and energy storage.

Starting from the earlier methodologies (pinch methodology, exergy analysis), this book tries to introduce the methodology and tools necessary for the design of energy-efficient and flexible systems.

I.1. Systemic constraints

Since the industrial revolution, the world population has grown exponentially from 1 billion around 1850 to 7.4 billion today (Figure I.1). It increases at a staggering rate of 80 million people per year. This expansion can be easily associated with technological innovations, in particular, with the rise and mastering of energy provided by fossil fuels.

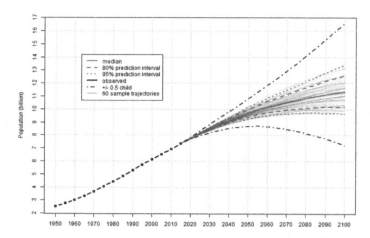

Figure I.1. *World's population since the industrial evolution and its possible variations [UNI 15]. For a color version of this figure, see www.iste.co.uk/zoughaib/pinch.zip*

It resulted in a tremendous and unprecedented period of prosperity for human civilization (Figure I.2). The exponential growth of the world's Gross Domestic Product (GDP) from the 1850s is evidence of this.

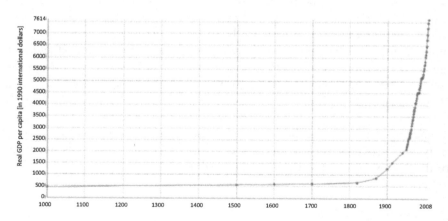

Figure I.2. *World's GDP over the last 1000 years[1]*

However, this population and economic growth has put a great and steady pressure on the environment since more and more natural resources are

1 https://ourworldindata.org/gdp-growth-over-the-last-centuries/.

consumed with an increase in waste and pollutions as a result. Indeed, in the prevailing linear economic model, resources are extracted and transformed to produce goods and services which are then consumed, producing waste that is more or less discarded back into the environment.

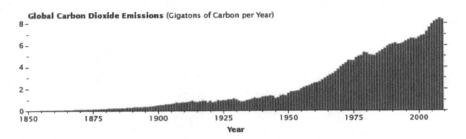

Figure I.3. *CO_2 emissions since the industrial revolution[2]*

Unfortunately, human activities have been profoundly linked to an increase in greenhouse gas emissions and pollution around the world (Figure I.3). This has resulted in dramatic changes in the climate and environment with significant consequences on human lives (natural disasters, health issues, political conflicts), which may worsen if the current conditions do not improve. Knowing that in order to limit global warming to 2°C, a maximum of CO_2 emissions of approximately 3,000 gigatonnes (Gt), of which two-thirds have already been emitted [IEA 15a], should not be exceeded. This suggests that the current economic model is not sustainable for the planet nor its inhabitants. A steadily increasing proportion of the population in developing countries (basically countries which are not OECD[3] members), particularly in China and India where one-third of the world's population lives, are demanding higher living and working standards, driven by the globalization of markets, information and culture. Therefore, building infrastructures is necessary to provide a more reliable access to food, water and energy as well as all goods and services that the modern economy of developed countries offers to its inhabitants. For that purpose, their industrial sector will consume more and more natural resources and energy (in particular fossil fuels, as shown in Figure I.4) to meet the growing demands.

2 Graph by Robert Simmon, using data from the Carbon Dioxide Information Analysis Center.
3 Organization for Economic Co-operation and Development.

The world population is expected to reach 9 billion people around 2050 (Figure I.1). More and more people will live in cities (by 2050, nearly 70% of the world population will live in urban areas [OEC 14]) and will have access to all goods that the economy can offer (cars, computers, Internet, smart phones, etc.).

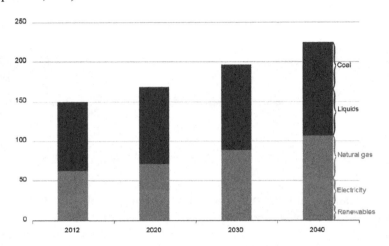

Figure I.4. *Non-OECD industrial sector energy consumption forecasts (×1015 Btu) [DOE 16]. For a color version of this figure, see www.iste.co.uk/zoughaib/pinch.zip*

This will create a strong competition for land (between agriculture, industry and housing) and resources (especially in water stress regions between drinking water for people and water necessary for manufacturing goods). Therefore, new ways of producing and consuming must be found to sustain the rising population and its needs while avoiding environmental damages and resource scarcity.

I.2. Energy consumption and production

Currently, the world consumes on average 150,000 TWh of primary energy per year. It has more than doubled over the last 40 years (Figure I.5). Fossil fuels (coal, oil, gas) represent the main energy source with 78% of the world primary energy consumption. They are also the main source of greenhouse gas (GHG) emissions.

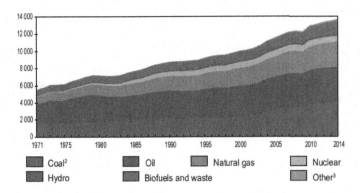

Figure I.5. *World's total primary energy consumption from 1971 to 2014 by fuel (Mtoe) [IEA 16]. For a color version of this figure, see www.iste.co.uk/zoughaib/pinch.zip*

The world final energy consumption is roughly divided between residential (20%), transport (25%) and industry (55%), mainly provided by fossil fuels [IEA 16]. As shown in Figure I.6, industrial activities account for more than 50% of world GHG emissions (32 Gt in 2014), and its share has been growing rapidly in non-OECD countries (mostly China and India) related to their economic growth. In OECD countries, relocation of industrial activities as well as measures to improve the energy efficiency have helped control the level of emissions in the last 25 years. However, the global GHG emissions have increased by about 50%.

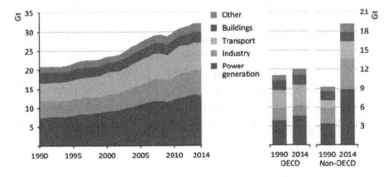

Figure I.6. CO_2 *emissions between 1990 and 2014 [IEA 15a]. For a color version of this figure, see www.iste.co.uk/zoughaib/pinch.zip*

By 2050, the world economy is expected to be four times larger than it is today [OEC 14]. Energy needs will be 80% higher than today if no changes are implemented by then, whether on technical, cultural, political, economic or social levels. Even though the part of renewable energies in the world energy mix is likely to grow faster than any other types of energy source, fossil fuels will remain the most important energy source (Figure I.7).

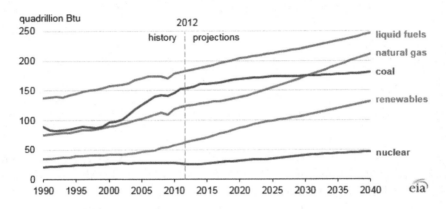

Figure I.7. *Projection of world's final energy consumption by fuel by 2040 (×1015 Btu) [DOE 16]. For a color version of this figure, see www.iste.co.uk/zoughaib/pinch.zip*

Even though their integration and management on the electrical grid remains a complex problem to be solved due to their intermittent nature, renewable energies are the future of world energy supply. One complementary solution that can help answer current environmental and economic issues would be to reduce energy needs by radically improving the efficiency of existing urban and industrial processes. Indeed, energy efficiency aims at reducing energy requirements to produce the same output. In Figure I.8, it represents approximately 40% of the total potential for GHG emission reduction. In particular, it accounts for more than two-thirds of the industry potential.

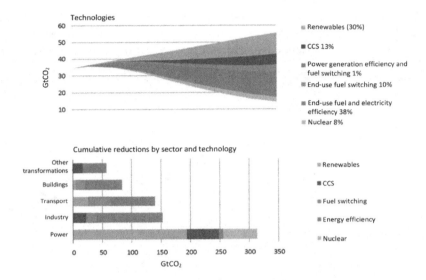

Figure I.8. *Technical solutions to limit CO_2 emissions and their potential [IEA 15b]. For a color version of this figure, see www.iste.co.uk/zoughaib/pinch.zip*

Energy efficiency has various economic advantages at different scales. For a company, it can increase product profitability and reduce the dependency towards energy suppliers. For a country, it helps avoid energy importations and controls its trade balance, and will have deep political implications for its geopolitical strategies and budget allocations (see Figure I.9).

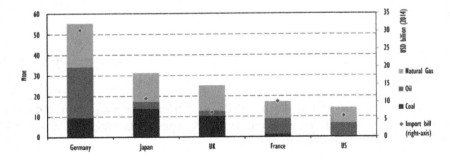

Figure I.9. *Avoided volume and value of energy imports from efficiency investments [IEA 15c]. For a color version of this figure, see www.iste.co.uk/zoughaib/pinch.zip*

As shown previously, fossil fuels are likely to remain the main energy source for this century. However, since the beginning of their use, oil and gas markets have experienced significant price volatility (Figure I.10).

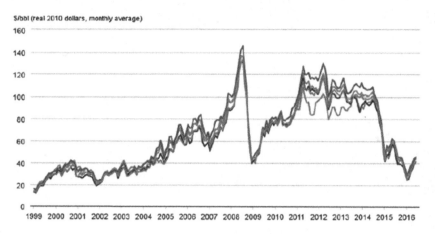

Figure I.10. *Crude oil price variations since the beginning of the century[4]. For a color version of this figure, see www.iste.co.uk/zoughaib/pinch.zip*

The causes of these variations are often rooted in political turmoil (wars in Iraq, Syria and Libya), world economic crises (sub-prime crisis in 2008) or technical uncertainties (supposed oil peak occurrence, US shale gas potential). These variations greatly influence social and environmental policies as well as profitability of investments. Therefore, reducing energy needs can and will limit the economic impact of swift variations in fuel prices that are likely to occur in the decades to come.

I.3. Current and future regulations – constraints on industrial actors

Political actions have been taken at different scales (international, regional or national) to improve resource use efficiency and limit the environmental footprint of human activities.

In the early 1990s, the Kyoto Protocol set targets to signing partners to reduce their CO_2 emission levels by 2010. In March 2007, the European

---

4 Bloomberg, Thomas Reuters. Published by U.S. Energy Information Administration.

Union adopted new environmental targets even more ambitious than that of the Kyoto Protocol for 2020[5]: 20% in renewable energies in its energy mix, 20% of energy consumption saved by increasing energy efficiency, 20% reduction of GHG emissions based on 1990 levels. Recently, countries met at the COP21 conference and agreed on new objectives for the next 30 years.

In order to reach the objectives in terms of energy efficiency, France decided to base its strategy on energy and fuel suppliers such as EDF, TOTAL or ENGIE [CGE 14]. These companies are expected to reduce their energy consumption or their clients' by a given amount set over a period of time. For instance, the objective was set at 345 TWh cumac (cumulated and actualized) over the period of 2011–2014. EDF was in charge of 40% of the total[6]. In case the target is not met, the companies would be compelled to pay fines.

Therefore, these companies are looking for new and innovative ways to improve the energy efficiency of their clients, creating a market for energy efficiency certificate that helps improve the financial feasibility of radical energy efficiency measures.

These regulations push industrial actors to continue innovating to keep ahead of their competitors. One of the most promising solutions is to improve the efficiency of existing processes. One way to do so is to consider effluents no longer as liabilities but as energy resources with great potential. This situation favors, in addition to the traditional energy integration through heat exchanger's network, innovative energy conversion systems (heat pumps, organic Rankine cycles, etc.).

In summary, the abundant and easy access to energy sources, thanks to fossil fuels, started an era in human history that led through unprecedented economic and demographic growth. Unfortunately, the massive consumption of fossil fuels, as well as natural resources, has had major consequences on the environment. Now, since the economic and demographic trends are more than likely to keep moving in the same direction and an important part of the world's population will need and want easy access to utilities and commodities, there is a critical need to shift our production and consumption habits.

---

5 ec.europa.eu/clima/policies/strategies/2020/.
6 travaux.edf.fr/construction-et-renovation/demarches-on-vous-guide/les-certificats-d-economies-denergie.

Moreover, focusing on the industry, companies must face new and necessary environmental regulations acting on their economic performances. Therefore, technical solutions must be designed to be cost-effective and meet modern ecological standards.

I.4. Book motivations and organization

In the last century, the thermodynamic community introduced concepts derived from the first and second laws of thermodynamics, helping to improve energy conversion systems. The term "exergy" was introduced in 1956 by Zoran Rant (1904–1972) by using the Greek *ex* and *ergon* meaning "from work". Exergy analysis and exergy-guided system optimization has since then been developing within this community.

In parallel, in the chemical engineering community, heat integration methodology developments started with the pinch methodology proposed by Lindhoff and Hindmarsh [LIN 83]. These developments were impulsed by the oil crisis in the early 1970s caused by geopolitical tensions in the Middle East. The implementation of this methodology in the oil and gas industry helped save energy and reduced the dependency on the oil-producing countries. Since then, many developments have been proposed, moving from the graphical original methodology to more complex mathematical programming methods.

Tackling the challenges that our century is facing asks for performing methodologies leading to radically higher energy-efficient solutions and innovative technologies for energy conversion. Heat integration and exergy analysis have to be coupled to provide systemic and systematic tools for designing energy-efficient and flexible processes.

This book reports the recent methodology developments introducing exergy concept to the heat integration techniques.

These formulations allow us to systematically assess the heat recovery options through:

– process modification;

– heat exchanger networks;

– thermodynamical energy conversion systems such as heat pumps, organic Rankine cycles and absorption heat pumps;

– heat storage.

The system architecture derived from the formulated problems defines operating conditions for energy conversion systems that have to be designed and optimized using exergy analysis-driven tools.

In addition to open literature works, this book details original developments where exergy consumption is introduced as an objective function to minimize, in mathematical programming, models for both continuous and batch processes. Most of these developments were done in the Center for Energy Efficiency of Systems in Mines ParisTech and reported in PhD dissertations and published articles.

The whole methodology is implemented in the open-source CERES[7] platform.

---

7 http://www.club-ceres.eu/.

Energy Integration of Continuous Processes: From Pinch Analysis to Hybrid Exergy/Pinch Analysis

1.1. Pinch analysis

1.1.1. *Basis of pinch analysis*

Process integration methodologies were developed in the 1980s, mainly in the oil and gas industry. These methodologies, including the most famous one, pinch analysis, allowed energy saving potentials through internal heat recovery, leading to drastic reduction in hot and cold utility use.

Linnhoff and Hindmarsh [LIN 83] introduced the pinch method. It is based on the process description as streams to be heated or cooled. The analysis of these streams, in addition to the one forming the utilities, by considering the heating and cooling power simultaneously with their temperature level, gives the pinch analysis.

The pinch analysis methodology is essentially graphical. Its main strength is that it allows us to rapidly visualize the streams of a process and deduce the potential heat recovery by internal heat exchange. Once the internal heat exchange opportunities are identified, the pinch methodology makes it possible to define directly to the Minimum Energy Requirement (MER). This is the minimum heating and cooling power to be provided by the utilities if all the internal heat exchange potential is exploited.

1.1.2. *Building the heat cascade*

We call a source (or hot stream) a fluid that needs to be cooled and a sink (or cold stream) a fluid that needs to be heated within the process. These needs can be described by the mass flow rate, the specific heat capacity and the temperatures at the inlet and outlet of the unitary operation using the following equation:

$$\Delta H = \dot{m} \times Cp \times \Delta T = CP \times \Delta T \qquad\qquad [1.1]$$

where:

ΔH is the enthalpy flux change through the unitary operation;

$\dot{m}$ is the fluid mass flow rate;

Cp is the specific heat capacity;

CP is the heat capacity flux;

ΔT is the temperature change through the unitary operation ($T_{in} - T_{out}$).

In this formulation, it is assumed that CP is independent of temperature or its variation within the stream temperature range is neglected. Phase changes may be represented in a similar way by considering a very small temperature change (arbitrarily chosen for pure fluids or azeotropic mixtures) or the phase change temperature change (non-azeotropic mixtures) and an equivalent Cp defined as the ratio of the latent heat and the considered temperature difference.

Let us consider the simple four-stream example in Table 1.1, initially reported in [KEM 07].

Stream number and type	CP (KW/K)	Tin (°C)	Tout (°C)
C1 Cold	2	20	135
H1 Hot	3	170	60
C2 Cold	4	80	140
H2 Hot	1.5	150	30

Table 1.1. *Streams and temperatures*

For a heat exchange to be feasible between a hot and a cold stream; a minimal temperature difference ΔT_{min} between these streams is required.

This temperature difference is associated with a finite heat exchange area and a heat exchanger technology.

In order to represent all the streams on a single temperature scale (for both hot and cold streams), we have to consider the heat exchanger wall temperature. Usually, we consider an equivalent heat transfer resistance for each side of the heat exchanger. By this assumption, a unique temperature scale may be obtained by shifting down the hot streams by ½ ΔTmin and by shifting up by ½ ΔTmin the cold streams.

To illustrate this, Table 1.2 presents the shifted stream temperatures of the simple four-stream example given in Table 1.1, considering a ΔTmin of 10°C.

Stream number and type	CP (KW/K)	Tin (°C) (shifted)	Tout (°C) (shifted)
C1 Cold	2	25	140
H1 Hot	3	165	55
C2 Cold	4	85	145
H2 Hot	1.5	145	25

Table 1.2. *Streams and shifted temperatures*

Thanks to the shifted temperature scale, hereafter, hot and cold streams may be represented as shown in Figure 1.1.

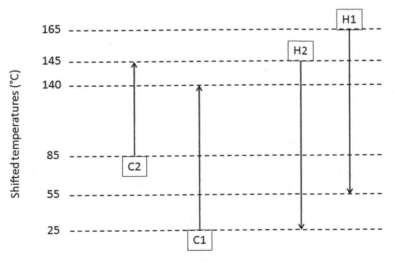

Figure 1.1. *Stream representation using the shifted scale*

The qualitative representation in Figure 1.1 shows the starting point of a stream (represented by the box with the label) and the end point of the stream (represented by the arrow). This graph helps identify temperature intervals where hot and cold streams coexist. In these intervals, we can expect hot and cold streams to exchange heat. This heat exchange potential is identified; however, it is not possible with such a representation to determine the heat power that may be exchanged.

Therefore, we need to determine the enthalpy heat balance over each interval. This heat balance allows us to determine whether the streams in the interval present a heat surplus or a heat deficit. Performing this heat balance over all the temperature intervals in the scale (starting from the highest temperature interval) is called the heat cascade, since the heat surplus of a higher interval may be cascaded to the one below.

As shown in Table 1.3, in each temperature interval, the surplus or deficit heat is calculated. The surplus of the temperature interval i may be used to compensate a deficit in the interval $i + 1$ (since the interval i is at a higher temperature level than the interval $i + 1$). Such a cascade avoids both a cold utility evacuating the surplus of the interval i and a hot utility supplying the deficit in the interval $i + 1$.

Interval number	ΔT_i in°C	$\sum CP_{hot} - \sum CP_{cold}$ (KW/°C)	ΔH_i in KW
1 (165–145°C)	20	3.0	60
2 (145–140°C)	5	0.5	2.5
3 (145–85°C)	55	−1.5	−82.5
4 (85–55°C)	30	2.5	75
5 (55–25°C)	30	−0.5	−15

Table 1.3. *Surplus or deficit heat per interval*

If no heat is supplied from the utility at 165°C, when doing the heat balance over the cascade (cascading surplus and deficit from higher intervals to the lower ones), we can determine the cumulated net heat balance at each interval as shown in Table 1.4.

Temperature (°C)	ΔHi in KW	ΔH net in KW
165		0
	60	
145		60
	2.5	
140		62.5
	−82.5	
85		−20
	75	
55		55
	−15	
25		40

Table 1.4. *Net cumulated heat cascade*

The net cumulated heat balance at a certain temperature level cannot be negative (which means that the heat is transferred from cold streams to hot streams). In order to make the cascade thermodynamically feasible, we need to provide heat in a way to eliminate all negative cumulated heat balances. In our example, we need to provide 20 kW starting from the highest temperature level as shown in Table 1.5.

Temperature (°C)	ΔHi in KW	ΔHnet in KW
165		20
	60	
145		80
	2.5	
140		82.5
	−82.5	
85		0
	75	
55		75
	−15	
25		60

Table 1.5. *Net cumulated heat cascade with heat addition*

The construction of the heat cascade therefore allows us to determine the heat needed for the process once all the heat surplus has been used for the cold stream heating. This heating power is the minimum energy requirement (MER) provided by the hot utility. It can be read easily at the top of the heat cascade. At the lowest point of the cascade, we can see that all the surplus has not be used and heat has to be discarded by the cold utility. This cooling power is called the cold MER. We may also remark that the net heat in the cascade passes by 0 at a certain temperature (in the example, this temperature is at 85°C). This point is defined as the pinch point of the process.

1.1.3. *Composite curves (CC) and grand composite curve (GCC)*

Composite curves are used to simultaneously visualize hot streams, cold streams and the heat transfer potential between them on the same graph. It is a graphical technique for determining the heat cascade. In each temperature interval (i in the hot stream scale and j in the cold stream scale) (we use the original temperature scale), the total heat power is calculated separately for the hot and cold streams (equations [1.2] and [1.3]). Then, the graph (Q,T) is plotted for both hot and cold streams:

$$\Delta \dot{H}_{hot,i} = \sum CP_{hot,i} \times \Delta T_i \qquad [1.2]$$

$$\Delta \dot{H}_{cold,j} = \sum CP_{cold,j} \times \Delta T_j \qquad [1.3]$$

The composite curves represent both the heat source and heat sink profiles. The superposition of the source (hot CC) and sink (cold CC) curves allows us to graphically retrieve the MER and the MER cold as found numerically in the heat cascade.

Note that the hot CC has to be on top of the cold CC in order to have a feasible heat transfer. The minimum temperature difference (ΔTmin) between the two curves guarantees a technically and economically feasible heat exchanger network. We start plotting the hot CC starting from the origin abscissa. The cold CC is plotted such that the minimal distance with the hot CC is ΔTmin.

The composite curves of our example are plotted in Figure 1.2.

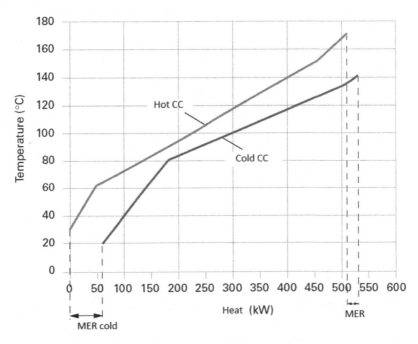

Figure 1.2. *Composite curves. For a color version of this figure, see www.iste.co.uk/zoughaib/pinch.zip*

From this graph, we can retrieve the pinch point and the values of the MER and MER cold that we already determined in Table 1.5. The pinch point is the temperature where the distance between the CC curves is ΔTmin. The MER is the final abscissa difference of both curves, whereas the MER cold is the initial abscissa difference.

It is also practical to plot shifted composite curves (using the shifted temperature scale). This is obtained as in Figure 1.3 by translating down the hot CC by ΔTmin/2 and translating up the cold CC by ΔTmin/2. In this representation, the curves are in contact at the pinch point.

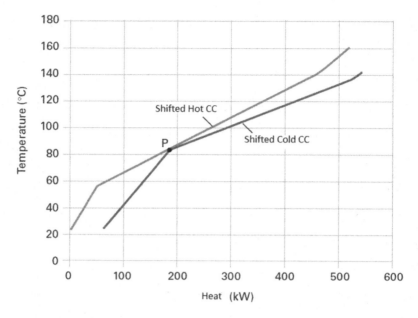

Figure 1.3. *Shifted composite curves. For a color version of this figure, see www.iste.co.uk/zoughaib/pinch.zip*

A complementary representation to the composites curves is the grand composite curve (GCC). It is the graphical representation of the heat cascade. It can be directly determined from the numerical method; directly obtaining the coordinates of the curve from Table 1.5. It is also possible to retrieve these coordinates graphically from Figure 1.3. Indeed, for each temperature in Figure 1.3, we can determine the net heat power as the abscissa difference between the hot and cold CC curves.

The GCC of our example is plotted in Figure 1.4.

On the GCC, we can identify two zones. Above the pinch point P, there is an endothermic zone with a heat deficit. Below the point P, an exothermic zone where cooling is needed.

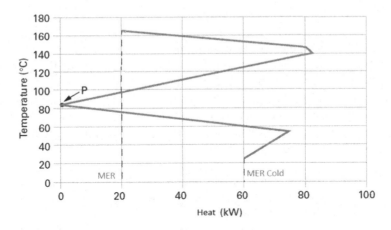

Figure 1.4. *The grand composite curve (GCC). For a color version of this figure, see www.iste.co.uk/zoughaib/pinch.zip*

1.1.4. *Pinch rules*

The separation of the problem into two independent zones separated by the pinch allows us to define three rules. Respecting these rules helps us to design a heat exchanger's network and select adapted utilities, therefore targeting a high energy efficiency level and approaching the minimal energy requirement.

These rules are the following:

– no heat exchange crossing the pinch point;

– no hot utility below the pinch point;

– no cold utility above the pinch point.

Indeed, the heat cascade is determined such that the surplus heat at a temperature interval may be useful in the next one. At the pinch point, we have the highest cumulated heat deficit in the process. Therefore, if we make a heat exchange crossing the pinch, we use a heat surplus located above the pinch to provide heat to a stream below the pinch. This heat is no more available above the pinch which causes, as a result, a higher heat deficit at the pinch point. As a result, the hot utility and the cold utility increase accordingly.

1.1.5. *Utility targeting using the GCC*

The GCC shows heating and cooling needs at each temperature level. The construction of this curve (by making the heat cascade) gives the cumulative heating needs by moving from the pinch at the higher temperatures. In the same manner, cooling needs are cumulated by starting from the pinch point at the lower temperatures. Therefore, graphically, we can determine the exact heating and cooling power to be provided at each temperature level without causing pinch point position modification.

If the GCC presents slopes changing sign in a zone (above or below the pinch); such a slope sign change indicates, for instance in the zone above pinch, a net cooling need. The heat cascade allows us to demonstrate that this cooling load is cascaded to provide a heating load at a lower temperature. This forms what we call "self-sufficient pockets" as shown in Figure 1.5 for our example.

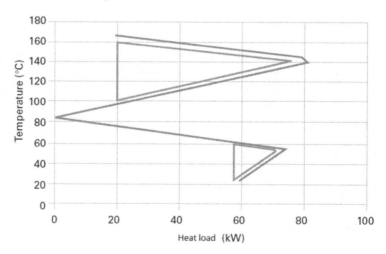

Figure 1.5. *GCC and self-sufficient pockets. For a color version of this figure, see www.iste.co.uk/zoughaib/pinch.zip*

The self-sufficient pockets are generally neither heated nor cooled by utilities, since such operation affects the pinch point position. In some special cases, as shown by [THI 13], it is interesting to heat and cool these pockets without modifying the heat equilibrium. This will be very useful when integrating thermodynamic conversion systems, as shown in section 1.2.

This rule allows us to size utilities and select their temperature as shown in Figure 1.6(a) and (b).

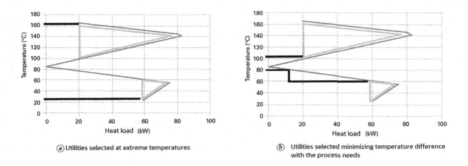

(a) Utilities selected at extreme temperatures

(b) Utilities selected minimizing temperature difference with the process needs

Figure 1.6. *Utility placement using GCC. For a color version of this figure, see www.iste.co.uk/zoughaib/pinch.zip*

Figure 1.6 shows two situations:

– in Figure 1.6(a), standard utilities are used (medium temperature steam for heating and cooling water). These utilities provide the minimum energy requirement as determined by the heat cascade. However, a simple observation of the figure shows a high temperature difference between the process (dark gray curve) and the utility (black curve). The second law of thermodynamics means we expect a high level of exergy destruction;

– in Figure 1.6(b), heat is provided at a temperature close to 100°C (since the highest part of the curve is a self-sufficient pocket). Cooling is staged at two temperature levels (80°C and 60°C). Graphically, it is possible to see that the temperature difference between the utility curve and the process curve is reduced compared to the situation in Figure 1.6(a) indicating less exergy destruction.

Hence, in Figure 1.6(b), low temperature heat is used for heating the process. This heat load (at this temperature level) may be recovered from a neighbor process or from a CHP (Combined Heating and Power) system. On the other hand, the cooling of two temperature levels creates opportunities for valorizing this heat (since at a higher temperature, it has a higher value (exergy content)). This heat may be sold to a neighbor process, converted into electricity using an organic Rankine cycle or reintegrated into the process through a heat pump.

Combining the exergy concept with pinch methodology not only allows us to limit the process' heating and cooling needs by cascading energy, but also to design utilities and conversion systems around the process, which leads to minimizing exergy requirement instead of energy requirement.

1.2. Exergy-based methodology for thermodynamic system systematic integration

The last step when integrating the process using a pinch study is the Heat Exchangers Network (HEN) design. This step, as well as available methodologies and models, is detailed in section 1.3. However, although methodologies exist to design HEN, as a manual methodology or in different ways of programming (MILP, NLP, MINLP, EA, SA, etc.), it is the most complex and time consuming. Input data have to be set very precisely especially when it comes to utility and conversion system options. In order to minimize the global cost (HEN and utilities), several utility options have to be considered.

If too many utilities are available in the set, HEN models may crash because of complexity. Moreover, relevant utilities that are not submitted will never be chosen.

As shown in the previous section, exergy-driven reasoning allows us to guide the design of the utility system and presents the opportunity to consider heat conversion systems that improve the overall energy efficiency.

Respecting the pinch rules and the second law of thermodynamics helps to determine the zones where different possible heat conversion systems may be integrated, as shown in Figure 1.7.

Figure 1.7 shows that the GCC is divided into three zones delimited by the pinch point and the ambient temperature. The ambient temperature is introduced by the second law while the pinch temperature by the pinch methodology.

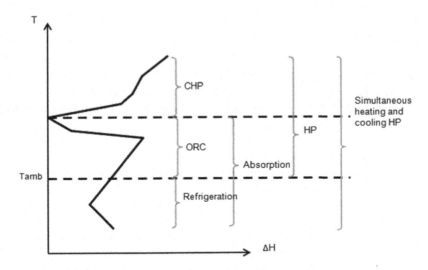

Figure 1.7. *Zones for integrating heat conversion systems*

In Figure 1.7, several systems and their respective integration zones are defined:

– CHP: Combined Heating and Power systems are hot utilities valorizing combustion exergy in both forms: power and heat. It is integrated above the pinch;

– ORC: Organic Rankine Cycles convert excess heat in the process into power. It is a cold utility. It should be integrated below the pinch. The ORC needs a heat sink for heat rejection, therefore it should be integrated above the ambient temperature;

– Absorption refrigeration systems: these tri-thermal cycles use heat to regenerate an absorbent that acts as a chemical compressor for a refrigerant. The refrigerant condenses at a heat sink (ambient temperature) and evaporates by absorbing heat at a heat source (below ambient temperature). The refrigerant is finally absorbed by the absorbent, releasing heat to a heat sink (ambient temperature). Therefore, this system acts as two cold utilities: one below the pinch point and above the ambient temperature, and one below the ambient temperature;

– HP: heat pumps using ambient temperature as a heat source may be used as hot utilities. In this case, we pump heat from the ambience and release it to the process above the pinch. We call them HPUt (heat pump utility);

– Simultaneous heating and cooling heat pumps: the heat source of the heat pump may be the process. In this case, the heat pump acts as both a cold utility and a hot utility. In this case, it has to be integrated at both sides of the pinch. This rule will be generalized to include the potential pinch point as shown hereafter. We call them HPPr (HP processes).

Manual and expert application of such reasoning is feasible but limited. Therefore, the use of mathematical programming allows a systematic analysis with more precise results.

The present model was developed by Thibault *et al.* [THI 15]. The model makes automatic preselection and pre-design of utilities in one step. This MILP (Mixed Integer Linear Programming) algorithm is based on the GCC utilization and a simplified exergy definition. Independently from their cost, utilities proposed by the present algorithm will be chosen according to their energy efficiency, based on exergy criteria. The MILP algorithm has the following features:

– defines the number of utilities of each type;

– makes a useful step between data extraction and HEN design;

– tests various, fitted combinations of utilities;

– substitutes the usual utility targeting step.

1.2.1. *Mathematical model*

1.2.1.1. *Input data*

The algorithm is based on the GCC. As the GCC represents the net heat load needed for each temperature level, the implicit assumption is that possible direct heat exchanges have been implemented. The remaining heat needs have to be provided by external utilities. In addition to basic utilities (e.g. boilers), the algorithm pre-designs advanced thermodynamic utilities (heat conversion systems), like heat pumps, ORC or CHP units.

The algorithm uses the GCC curve as input data. The vertical axis, the temperature range, is subdivided into several sections, delimited by the Main Pinch Point (MPP) and minimum heat load, hereafter called the Potential

Pinch Point (PPP) (Figure 1.8). This defines a number of zones (Z), each of them delimited by two pinch points (main or potential). At each PPP, self-sufficient pockets appear: when there is the same heat load at two different temperatures, an increase followed by a decrease of heat load required generates a self-sufficient pocket.

In each zone z, the angular points of the GCC are considered to create a first set of temperature intervals. These correspond to incoming or depleted heat flow. Moreover, a maximal temperature step is introduced as a parameter. Any higher temperature interval will be halved until it respects this parameter. It is a key parameter to balance speed and accuracy of the algorithm that should be applied to the process studied.

Once the temperature intervals are defined, the parameter S_z defines the number of temperature segments in each zone z. Each point is therefore described by two parameters, $T_{z,k}$ and $Q_{z,k}$, respectively temperature and heat load, with $z \in [1, Z] \, et \, k \in [1, S_z]$.

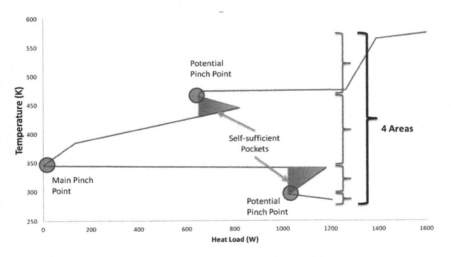

Figure 1.8. *Self-sufficient pockets and pinch points [THI 13]. For a color version of this figure, see www.iste.co.uk/zoughaib/pinch.zip*

1.2.1.1.1. Conversion systems modeling

The aim of this algorithm is to pre-design utilities that fit the best process for the exergy criteria. At this preliminary step, simplified models based on thermodynamic laws are developed, rather than consistent models.

Pre-designing utilities means finding the optimal operating temperature levels for heat conversion systems. Simple models are defined for each technology, based on ideal cycle efficiency and exergy efficiency (second law efficiency). Defined as the ratio between the real cycle performance and the ideal one, the exergy efficiency is commonly set between 0.5 and 0.6 (see Chapter 2). This value is considered as achievable for most of the technologies that are investigated.

Neither temperature levels nor heat load is known *a priori* for the utilities to be used. In the case of heat pumps, heat loads are linked to temperature with COP (coefficient of performance), which is temperature dependent; the problem could be nonlinear. To avoid nonlinearity, but also to drastically limit the number of utilities and ease post-processing, a linear formulation with Boolean variables is proposed. A very large number of utilities will be described, using each couple of elements of the discrete temperature axis (j in zone y and k in zone z such as T[y, j] < T[z, k]). For each utility and for each couple of temperature the coefficient of performance or the first law efficiency (COP$_{ref}$[y, j], COP$_{HPPr}$[y, j, z, k], COP$_{HPUt}$[y, j], Eff$_{ORC}$[y, j], Eff$_{Chp}$[j]) is precalculated with equations [1.4]–[1.8] before the optimization problem:

$$COPHPPr_{y,j,z,k} = \eta_{ex} * \frac{T_{z,k} + CondP}{\left(T_{z,k} + CondP\right) - \left(T_{y,j} - EvaP\right)} \qquad [1.4]$$

$$COPHPUt_{y,j} = \eta_{ex} * \frac{T_{y,j} + CondP}{\left(T_{y,j} + CondP\right) - \left(T_0 - EvaP\right)} \qquad [1.5]$$

$$COPref_{y,j} = \eta_{ex} * \frac{T_0 + CondP}{\left(T_0 + CondP\right) - \left(T_{y,j} - EvaP\right)} \qquad [1.6]$$

$$EffORC_{y,j} = \eta_{ex} * \left(1 - \frac{T_0 + EvaP}{\left(T_{y,j} - CondP\right)}\right) \qquad [1.7]$$

$$EffChp_j = \eta_{ex} * \left(1 - \frac{T_{z,j} + EvaP}{\left(T_f - CondP\right)}\right) \qquad [1.8]$$

where:

– EvaP and CondP are the minimum temperature difference at the evaporator and condenser respectively;

– η_{ex} is the exergy efficiency or the second law efficiency of the heat conversion systems;

– T_0 is the ambient temperature that is also considered as the reference temperature for the exergy calculation;

– T_f is the flame temperature when converting chemical exergy of fuel into heat.

Two variables, one Boolean and one continuous, are used to describe respectively the presence of the utility (BoolChp$_k$, BoolRef$_{z,k}$, BoolORC$_{z,k}$, BoolHPPr$_{y,j,z,k}$, BoolHPUt$_{z,k}$) and the fraction of the available heat load used (FChp$_k$, FRef$_{z,k}$, FORC$_{z,k}$, FHPPr$_{y,j,z,k}$, FHPUt$_{z,k}$). The use of the Boolean variable allows us to count utilities and to set a maximum number of utilities to pre-design as a constraint. As utility power cannot exceed available or needed heat load at a temperature, equations [1.9]–[1.13] both limit the factor and nullify it if the utility is absent:

$$FChp_k \leq BoolChp_k \qquad\qquad [1.9]$$

$$FRef_{z,k} \leq BoolRef_{z,k} \qquad\qquad [1.10]$$

$$FORC_{z,k} \leq BoolORC_{z,k} \qquad\qquad [1.11]$$

$$FHPPr_{y,j,z,k} \leq BoolHPPr_{y,j,z,k} \qquad\qquad [1.12]$$

$$FHPUt_{z,k} \leq BoolHPUt_{z,k} \qquad\qquad [1.13]$$

In order to avoid a very large number of utilities or small-sized utilities and to ease data post-processing, a maximum number of each technology is set up as a parameter:

$$\sum_{k \in [1, S_Z]} BoolChp_k < CHP_{max} \qquad\qquad [1.14]$$

$$\sum_{\substack{z \in [1, Z-1] \\ k \in [1, S_z]}} BoolRef_{z,k} < Ref_{max} \qquad\qquad [1.15]$$

$$\sum_{\substack{z \in [1, Z-1] \\ k \in [1, S_z]}} BoolORC_{z,k} < ORC_{max} \qquad [1.16]$$

$$\sum_{\substack{y, z \in [1, Z] \\ y < z}} \sum_{\substack{j \in [1, S_y] \\ k \in [1, S_z]}} BoolHPPr_{y,j,z,k} + \sum_{\substack{z \in [1, Z] \\ k \in [1, S_z]}} BoolHPUt_{z,k} < HP_{max} \qquad [1.17]$$

1.2.1.2. Energy balance

At each temperature level, a certain number of utilities may be used, each of them providing or taking a heat load. The different utility technologies can be divided into two sets:

– Process heat pumps (HP): they will exchange heat load between process streams, allowing thermal exchange from a cold stream to a warmer one. To be efficient, a process HP has to cross a pinch point. Evaporators of these HP will be set up in areas 1 to (Z−1), whereas condensers will be set up in areas 2 to Z;

– Utility heat pumps, chillers, ORCs and CHP units: they evacuate heat from a process to an external heat sink or provide heat from an external source. As chillers and ORC units remove heat, they have to be set below the MPP, whereas CHP units have to be set above the MPP.

Equations [1.18] and [1.19] evaluate the total heat load taken at $T_{y,j}$. Difference occurs between areas 1 to $Z-1$ and area Z, where only heat is provided. As we only consider the energy which is taken, just three types of utilities are concerned: process heat pumps, chillers and ORC units.

$$\forall y \in [1, Z-1], \forall j \in [1, S_y],$$

$$Pprel_{y,j} = \left(\left(\sum_{\substack{z \in [y+1, Z] \\ k \in S_z}} \left(FHPPr_{y,j,z,k} \right) \right) + FRef_{y,j} + FORC_{y,j} \right) Q_{y,j} \qquad [1.18]$$

$$\forall k \in [1, S_Z] \, Pprel_{Z,k} = 0 \qquad [1.19]$$

where:

Pprel$_{y,j}$ is heat extracted by utilities at T[y, j];

Q$_{y,j}$ is the net heat load at T[y, j] obtained from the heat cascade (GCC).

Below are the equations for the total heat provided at T$_{z,k}$. In the first zone, no heat will be provided, as there are remaining cooling needs. In intermediary areas, condensers of both process and utilities heat pumps can be set. In Zone Z, where there are only heating needs, CHP units will complete heat pumps to provide energy. The first law of thermodynamics and the definition of the coefficient of performance are used to link the heat taken by process heat pumps from T[y, j] with the heat released at T[z, k]:

$$\forall k \in [1, S_1], Papp_{1,k} = 0$$

$$\forall z \in [2, Z-1], \forall k \in [1, S_z], \tag{1.20}$$

$$Papp_{z,k} = \sum_{\substack{y \in [1,z-1] \\ j \in S_y}} \left(FHPPr_{y,j,z,k}\, Q_{y,j}\, \frac{COPHPPr_{y,j,z,k}}{COPHPPr_{y,j,z,k} - 1} \right) + FHPUt_{z,k}\, Q_{z,k}$$

$$\forall k \in [1, S_Z], \tag{1.21}$$

$$Papp_{Z,k} = \sum_{\substack{y \in [1,z-1] \\ j \in S_y}} \left(FHPPr_{y,j,z,k}\, Q_{y,j}\, \frac{COPHPPr_{y,j,z,k}}{COPHPPr_{y,j,z,k} - 1} \right) + \tag{1.22}$$

$$\left(FHPUt_{z,k} + FCHP_k \right) Q_{z,k}$$

where

Papp$_{z,k}$ is the heat provided at T[z, k];

Q$_{z,k}$ is the net heat load at T[z, k] obtained from the heat cascade (GCC).

1.2.1.3. *GCC's update*

The next step consists of rebuilding the GCC, updated so as to take into account the effect of utilities on process heat loads. In areas below the main pinch, adding a utility will reduce cooling needs below the temperature level of the utility. In the same way, in the upper areas, a utility providing heat load at any temperature implies a drop of heating needs all over the temperature equal to this heat load. Below the pinch, the net heat load of the updated GCC is built with equations [1.23] and [1.24] which updates the GCC above the pinch:

$$\forall y \in [1, Z-1], \forall j \in [1, S_y],$$

$$NHL_{y,j} = Q_{y,j} + \sum_{\substack{i\in[j,S_y] \\ j\in S_y}} \left(PApp_{y,i} - PPrel_{y,i} \right) + \sum_{\substack{y\neq Z-1 \\ z\in[y+1,Z-1] \\ k\in[1,S_z]}} \left(PApp_{z,k} - PPrel_{z,k} \right) \quad [1.23]$$

$$\forall j \in [1, S_Z], NHL_{Z,j} = Q_{Z,j} - \sum_{i\in[1..j]} PApp_{Z,i} \quad [1.24]$$

where

NHL$_{y,j}$ is the new heat load at T[y, j];

NHL$_{Z,j}$ is the new heat load in the last zone at T[Z, j].

A GCC can have more than one pinch point, but the creation of a new pinch point must not suppress the original MPP, which leads to an increase of MER. This is why the main constraint for the new GCC is:

$$\forall y \in [1, Z], \forall j \in [1, S_z], NHL_{y,j} \geq 0 \quad [1.25]$$

1.2.1.4. *Electricity production and consumption*

When all utilities are set up, the electricity consumption and production has to be calculated in order to evaluate overall exergy consumption. Heat pumps (process and utilities) and chillers consume electricity, whereas ORC and CHP units produce it. Only the net production is calculated, but more constraints could be added, as a production higher than the consumption (self-sufficiency of the process) or lower (in process usage but no sales of electricity). Local and total electricity consumptions are calculated with equations [1.26]–[1.28]:

$$\forall y \in [1, Z-1], \forall j \in [1, S_y],$$

$$Pelec_{y,j} = \sum_{\substack{z \in [y+1,Z] \\ k \in [1..S_z]}} \left(\frac{FHPPr_{y,j,z,k} * Q_{y,j}}{COPHPPr_{y,j,z,k} - 1} \right) + \frac{FHPUt_{y,j} * Q_{y,j}}{COPHPUt_{y,j}} + \frac{FRef_{y,j} * Q_{y,j}}{COPRef_{y,j} - 1} \quad [1.26]$$

$$\forall j \in [1, S_Z], Pelec_{Z,j} = FHPUt_{Z,j} * Q_{Z,j} * \frac{1}{COPHPUt_{Z,j}} \quad [1.27]$$

$$TEC = \sum_{z \in [1,Z]} \sum_{k \in [1,S_z]} Pelec_{z,k} \quad [1.28]$$

The electricity production is ensured by ORC units below the pinch and CHP units above the pinch. The total electricity production is:

$$TEP = \sum_{z \in [1,Z-1]} \sum_{k \in [1,S_z]} FORC_{z,k} * Q_{z,k} * EffORC_{z,k} +$$

$$\sum_{k \in [1,S_Z]} FCHP_k * Q_{Z,k} * \frac{EffCHP_k}{1 - EffCHP_k} \quad [1.29]$$

Because CHP units provide heat and produce electricity, they need a hot source to do this. The heat load necessary for all CHP units is:

$$Pprel_{CHP} = \sum_{k \in [1,S_Z]} FCHP_k * Q_{Z,k} * \frac{1}{1 - EffCHP_k} \quad [1.30]$$

1.2.1.5. Restriction on utility placement

As the problem uses binary variables, the number of possible solutions can dramatically increase. For instance, consider a GCC with two areas, 10 temperature points above the pinch and 20 below. A total of 200 different process heat pumps are potentially feasible. Depending on the number of heat pumps to be selected, the combinatory analysis gives a sky rocketing number of possibilities: 19,900 for two heat pumps to pre-design and more than 1.3 million for three heat pumps to pre-design.

As the algorithm aims to pre-design various numbers of different utility technologies, it has to be quick and accurate. The following constraints are also added to reduce the number of possibilities. The reference temperature T_0, considered as the temperature of the ambient heat sink temperature, is used to calculate exergy, but it will also define boundaries for some technologies as explained earlier. Below T_0, HPUt and ORC are forbidden:

$$if\ T_{z,k} < T_0, BoolHPUt_{z,k} = 0 \qquad\qquad [1.31]$$

$$if\ T_{z,k} < T_0, BoolORC_{z,k} = 0 \qquad\qquad [1.32]$$

In the same way, chillers are forbidden above the temperature T_0:

$$if\ T_{z,k} > T_0, BoolRef_{z,k} = 0 \qquad\qquad [1.33]$$

Then, in order to introduce a technology feasibility criterion for heat pumps, a new parameter is introduced: $T_{Condmax}$. This is the maximum temperature for setting condensers of heat pumps. Nowadays, heat pumps available off the shelf are limited to 100°C at condensers and this value rises to 130°C for some technologies under development. So, this parameter not only reduces the number of heat pumps to be tested, but also allows users to carry on different pre-design studies: at $T_{Condmax}$ = 100°C, existing technologies are tested and solutions are easily feasible, whereas at $T_{Condmax}$ = 130°C, if the gain is considerable compared to the previous case, new technologies could be considered and further developments would be mandatory. Finally, with higher $T_{Condmax}$, it is possible to identify R&D roadmaps for heat pump designers in some specific industrial fields.

These two equations correspond to $T_{Condmax}$ restriction:

$$if\ T_{z,k} > T_{Condmax}, BoolHPPr_{y,j,z,k} = 0 \qquad\qquad [1.34]$$

$$if\ T_{z,k} > T_{Condmax}, BoolHPUt_{z,k} = 0 \qquad\qquad [1.35]$$

1.2.1.6. *Objective function*

Finally, the objective function is total exergy consumption. It includes utility electricity consumption and production, heat source consumption of CHP units, and exergy needed for the remaining hot and cold MERs. The overall exergy consumption is calculated as follows:

$$FinalExergy = NHL_{1,1} * \eta_{ex} * \begin{cases} 0, & if\ T_{1,1} > T_0 \\ \dfrac{T_{1,1}}{T_0 - T_{1,1}}, & if\ T_{1,1} < T_0 \end{cases} +$$

$$\left(NHL_{Z,S_Z} + Pprel_{Chp} \right) * \frac{T_f - T_0}{T_f} + TEC - TEP$$

[1.36]

where NHL_{Z,S_Z} corresponds to the remaining hot MER and $NHL_{1,1}$ to the remaining cold MER.

1.2.2. *Case studies*

1.2.2.1. *Simple example*

Possibilities and utilization of this pre-design module for utilities is illustrated first with the four-stream example presented in Table 1.1.

The considered utilities depend on the profile of the GCC. In this case, heat pumps are relevant because the pinch point is low and simultaneous heating/cooling is needed. The general parameters of the algorithm for this case are summed up in Table 1.6. The discretization step leads to 102 temperature intervals. While the algorithm has been written in GLPK language, the solver CPlex (IBM) is being used to solve it.

Parameters	Values
Ambient Temperature (K)	288.15
Flame Temperature (K)	1,073.15
Pinch at Evaporators and condensers (K)	5
Maximal Temperature condensation for HP (K)	393.15
Maximal Temperature Discretization step (K)	2
Exergy Efficiency	60%

Table 1.6. *Preselection parameters*

The results are displayed with the hot and cold composite curves as well as with the Integrated Composite Curve, which represents the GCC surrounded on its left side by the utility curve. The temperature levels of heat pumps are guided by the exergy consumption minimization criteria, whereas the heat load is maximized to fit the utilities curve to GCC. Graphically, the aim is to obtain a utility curve close to the GCC which decreases the area between them. It means a decrease in exergy consumption and a better use of energy.

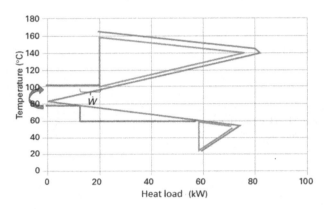

$\dot{W}$: Heat pump electrical power (kW)

Figure 1.9. *Preselected heat pump. For a color version of this figure, see www.iste.co.uk/zoughaib/pinch.zip*

Moreover, characteristics of the preselected heat pump are summarized in Table 1.7.

Parameter	Value
Heat pump heating capacity (kW)	20
Heat pump cooling capacity (kW)	17.25
Heat pump electrical power (kW)	2.75
Evaporation temperature (°C)	72.5
Condensation temperature (°C)	102.5
Remaining cooling load (kW)	42.75

Table 1.7. *Preselection result*

1.2.2.2. *Dairy case study*

The algorithm is tested with a dairy case described in the literature [BEC 12a] which includes singularities: many PPP over and below the MPP and multiple phase transitions. Simulations are run for 2 and 3 HPs.

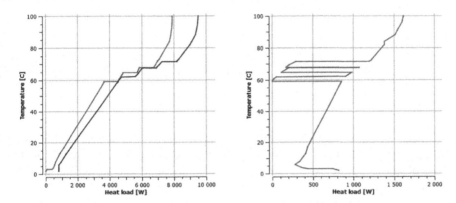

Figure 1.10. *Dairy process composites curves and grand composite curve [BEC 12a]. For a color version of this figure, see www.iste.co.uk/zoughaib/pinch.zip*

Ambient temperature is set at 25°C, corresponding to 288.15 K. ΔTmax, which is the maximum temperature difference between two discrete points of the GCC, is set to 4°C. As there is a possible HP between any pair of temperatures, a small temperature step will lead to an exponential number of feasible HP possibilities, according to combinatorial analysis. Exergy efficiency is set at 60%, pinch at evaporators EvaP is set at 2 K and pinch at condensers CondP is also 2 K.

Two calculations were run: the first one has to preselect 2 HPs and the second one 3 HPs. All the parameters are summarized in Table 1.8.

It is important to note the main difference between this study and the reference case study [BEC 12a]. In the latter, HPs are consistent models picked up among a set of 16 distinct HPs (four different refrigerants combined with four different compressors technologies), whereas in this model several hundred implementations of HPs are evaluated, but they are based on simplified models.

Parameter	Value
T0 (K)	288.15
ΔTmax	4
PACMAX	[2–3]
Exergy Efficiency	0.60
EvaP (K)	2
CondP (K)	2
Solver	GLPK

Table 1.8. *Algorithm parameters*

Figure 1.11 presents the result of MILP algorithm calculation for three HPs. It highlights that HP 3 is implemented to take advantage of the self-sufficient pocket by transferring additional heat into it, which enables HP 2 to efficiently transfer twice more energy from 58.9 to 71.5°C. In other words, HP 3 fills the self-sufficient pocket with energy and allows HP 2 to transfer more energy from it. Because of its lower COP, the MILP algorithm designs a smaller capacity for HP 1 (compared to the case with 2 HPs) which reduces the overall electricity consumption.

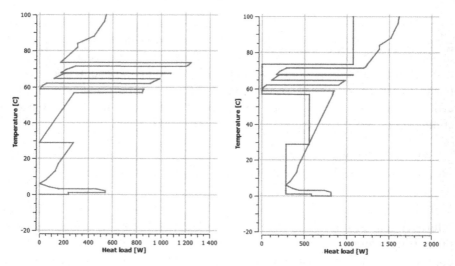

Figure 1.11. *GCC and Integrated Composite Curve (ICC) solutions (3 HPs case). For a color version of this figure, see www.iste.co.uk/zoughaib/pinch.zip*

Tables 1.9 and 1.10 summarize the results obtained for the two simulated cases and compare them to the best configuration obtained in Becker's study.

In the 2 HPs' solution, the delivery temperature level for HP 1 is different from Becker's study [BEC 12a], where HP 1 is an existing chiller, so it can be modified. However, its low COP is compensated by a higher heat load transfer and a large decrease of MER. In spite of higher electricity consumption, the total exergy consumption remains smaller than in Becker's case as summarized in Table 1.10.

	Evaporator		Condenser		
Solution	Temperature (°C)	Heat load (W)	Temperature (°C)	Heat load (W)	COP
2 HP's					
HP 1	0	537	73.5	830	2.8
HP 2	56.9	282	73.5	306	12.6
3 HP's					
HP 1	1	304	73.5	467	2.9
HP 2	56.9	559	73.5	607	12.6
HP 3	0	233	29	277	6.3
Becker's					
HP 1	−2	537	35	650	5.8
HP 2	56	615	76	668	12.6

Table 1.9. *HP characteristics*

When implementing 3 HPs, temperature levels of HP 2 and 3 are very close to Becker's HPs. On the other hand, Becker's algorithm is based on an economic criterion that certainly would eliminate HP 1. Even if the MER remains higher, implementing a third HP causes a decrease of exergy consumption and a drastic fall of electricity consumption, due to higher COPs.

In more complex processes, a large range of HP sets could be found by changing parameters. They are all optimal from an exergy point of view and are proposed as candidate utilities to the HEN algorithm implemented in CERES that find the best overall set of heat pumps, according to the economic criterion.

Final state	2 HPs	3 HPs	Becker's, 2012
Remaining MER (W)	478	545	975
Remaining MER$_{Cold}$ (W)	0	5	50
Electric consumption (W)	317	255	165
Exergy consumption (W)	652	637	852.5

Table 1.10. *Energy and exergy consumption*

1.3. Heat exchanger network synthesis

In industrial process synthesis, the heat exchanger network synthesis (HENS) is the essential step. Its objective is to design the heat exchangers required to achieve the heat integration and reach the MER determined by the pinch analysis. In this step, the objective is usually to design an economically optimal HEN which may in some cases lead to reconsidering the minimum temperature difference assumed in the pinch analysis step.

For the complete update on the HENS methodologies developed over the decades, the reader may consult the recent reviews on the topic, such as the papers [FUR 02, KLE 13]. Among the existing methodologies, the method based on pinch analysis is considered the most basic. For a detailed insight of this method, we may refer to [KEM 07]. It has been successfully applied in a large number of process synthesis projects across the world. However, it implies a manual calculation procedure, so it is difficult to be used for a complex system. In addition, there is no way to ensure that the solution found is the optimal one. These limitations require developing alternative methodologies, such as mathematical programming approaches.

In the mathematical programming approaches, the methods can be classified as either a sequential technique or a simultaneous technique. While the first technique is based on the strategy that divides the HENS problem into a number of sequences of calculation and generally uses a temperature partition, the simultaneous technique aims to find the optimal solution without decomposition of the problem. The simultaneous methods are based on mixed integer nonlinear programming (MINLP) formulations, which can raise a major difficulty related to the numerical resolution. Indeed, in the case of a non-convex problem, the solution found is probably the local optimum rather than the global one.

One of the first models using the sequential technique was presented in the paper [PAP 83]. Since this initial work, many models have been developed in order to design a more realistic heat exchanger network. Indeed, in almost all existing models, a number of assumptions have been made such as isothermal mixing, no split stream and no stream bypass, which allows us to reduce the complexity of the problem. Among the recent works, linear models have been presented in [JEZ 03] and then extended in [BAR 05]. Specifically, these models make it possible to approximate the heat exchanger areas, to implicitly determine flow rates in splits, to handle non-isothermal mixing and to permit multiple matches between two streams. A real plant layout associated with the space constraints can be also considered during the heat exchanger design [POU 14]. In addition, as the HENS is known as an NP-Hard problem [FUR 01], efforts in managing the computation time have been made [ANA 10, BEC 12b].

In this context, the model published in [TRA 15] which extends the model developed by Barbaro and Bagajewicz [BAR 05] in order to provide new functionalities to a designer and limit the required assumptions, is presented hereafter.

The model features are:

– It takes into account multiple heat exchanger (HEX) technologies. In real life, a great number of HEX technologies with different performances and costs exist. In addition, there is a number of constraints related to the use of HEX technologies due to the fact that the designer usually wants to impose a specific technology according to the properties of the streams. The proposed functionality allows the model to find the most appropriate technologies from an economic point of view while satisfying the constraints imposed by the designer.

– It considers flexible streams. In most of the developed methodologies throughout the years, only two types of stream are considered: process stream and utility stream [FUR 02, KEM 07]. For a process stream, the inlet and outlet temperatures and the mass flow rate are fixed. For a utility stream, only the temperatures are fixed, while the flow rate varies to satisfy the hot and cold requirements of the process streams. In most real processes a third type of stream can be identified, and it is referred to as a flexible stream. A flexible stream has an inlet temperature and a fixed mass flow rate, but its outlet temperature can vary. As an example, exhaust streams of a process, like hot fumes, are basically flexible streams because they can be either used for the heat recovery purpose or, under specific conditions, directly released into the environment. This model enables us to handle flexible streams. Specifically, the algorithm will determine the optimum outlet temperature and the associated heat exchanger network.

1.3.1. *Mathematical model*

1.3.1.1. *Base model*

This part sums up the base model which is presented in [BAR 05]. For the sake of simplicity, we present only the essential equations. Additional equations will be necessary in some specific conditions, for example when more than one match is permitted between two streams or when non-isothermal mixing is allowed. All these formulations are described in the original paper.

1.3.1.2. *Set definitions*

It is necessary to define a number of different sets that will be used through the model. First, a set of zones are defined, namely $Z = \{z|z$ is a heat transfer zone$\}$. The use of zones aims to separate the design into different sub-networks that are not interrelated, simplifying the problem complexity. As an example, if the designer wants to respect the rule of thumb based on the pinch method design (without heat transfer across the pinch point), two zones have to be defined (above and below the pinch temperature). This functionality is particularly interesting for a problem with a large number of streams, knowing that the HEN design is an NP-Hard problem [FUR 01], as the calculation time increases exponentially with the number of variables.

The following sets are used to identify hot streams, cold streams and utilities. Note that a set of streams includes process and utility streams:

$H^z = \{i|i$ is a hot stream present in zone z$\}$;

$C^z = \{j|j$ is a cold stream present in zone z$\}$;

$HU^z = \{i\,|i$ is a heating utility present in zone z$\}$ ($HU^z \subset H^z$);

$CU^z = \{j\,|j$ is a cooling utility present in zone z$\}$ ($CU^z \subset C^z$).

The temperature scale is divided, in each zone, into different intervals. This step allows the heat balances and the area calculations to be performed via linear equations. In addition, a shift of $\Delta T_{\min}$ is performed over all cold stream temperatures to guarantee heat transfer feasibility. We call T_m^U and T_m^L upper and lower temperatures of interval m. Moreover, different sets related to the temperature intervals need to be defined:

$M^z = \{m|m$ is a temperature interval in zone $z\}$;

$M_i^z = \{m|m$ is a temperature interval belonging to zone z, in which hot stream i is present$\}$;

$N_j^z = \{n|n$ is a temperature interval belonging to zone z, in which cold stream j is present$\}$;

$H_m^z = \{i|i$ is a hot stream present in temperature interval m in zone $z\}$;

$C_n^z = \{j|j$ is a cold stream present in temperature interval n in zone $z\}$.

Finally, the next sets allow the designer to set different constraints, according to his own preference:

$P = \{(i,j)|$ a heat exchange match between hot stream i and cold stream j is permitted$\}$;

$P_{im}^H = \{i|$ heat transfer from hot stream i at interval m to cold stream j is permitted$\}$;

$P_{jn}^C = \{j|$ heat transfer from hot stream i to cold stream j at interval n is permitted$\}$;

$NI^H = \{i|$ non-isothermal mixing is permitted for hot stream $i\}$;

$NI^C = \{j|$ non-isothermal mixing is permitted for cold stream $j\}$.

The sets P, P_{im}^H and P_{jn}^C are used to either permit or forbid specific heat exchange matches. The sets NI^H and NI^C allow the designer to specify whether non-isothermal mixing of stream splits is permitted.

1.3.1.3. Heat balance equations

The model is based on the transshipment/transportation (heat cascade) scheme. Figure 1.12 shows an example of a heat exchanger between hot stream i and cold stream j, in zone z. $q_{im,jn}^z$ represents heat transportation from interval m of hot stream i to interval n of cold stream j.

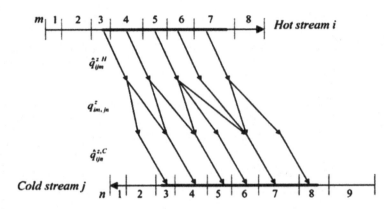

Figure 1.12. *Basic scheme of the transportation/ transshipment model [BAR 05]*

The heat balance equations state that the heat available on each hot stream or the heat demand of cold streams is equal to the heat transferred to the specific intervals. The following equations are used for hot utilities and hot process streams, respectively. The balance equations for the cold utility and cold process streams can be obtained in a similar way:

$$F_i^H \left(T_m^U - T_m^L \right) = \sum_{\substack{n \in M^z \\ T_n^L < T_m^U \\ i \in P_{jn}^C}} \sum_{\substack{j \in C_n^z \\ j \in P_{im}^H}} q_{im,jn}^z \quad z \in Z; m \in M^z; i \in H_m^z; i \in HU^z \qquad [1.37]$$

$$\Delta H_{im}^{z,H} = \sum_{\substack{n \in M^z \\ T_n^L < T_m^U \\ i \in P_{jn}^C}} \sum_{\substack{j \in C_n^z \\ j \in P_{im}^H}} q_{im,jn}^z \quad z \in Z; m \in M^z; i \in H_m^z; i \notin HU^z; i \notin NI^H \qquad [1.38]$$

where

$\Delta H_{im}^{z,H}$ is the enthalpy change for hot stream i at interval m in zone z;

F_i^H is the flow rate of hot utility stream i;

$q_{im,jn}^z$ is the heat transfer from hot stream i at interval m to cold stream j at interval n in zone z.

Note that for process streams, the enthalpy variations $\Delta H_{im}^{z,H}$ are considered as parameters and can be easily calculated because the flow rates and temperature intervals are known. On the contrary, for utilities the flow rates F_i^H are considered as variable and will be determined by the model.

The required heat transfer area for heat exchange between hot stream i and cold stream j is calculated as:

$$A_{ij}^z = \sum_{\substack{m \in M_i^z \\ j \in P_{im}^H ; i \in P_{jn}^C}} \sum_{\substack{n \in N_j^z ; T_n^L < T_m^U}} \frac{q_{im,jn}^z \left(h_{im} + h_{jn} \right)}{\Delta T_{mn}^{ML} \, h_{im} h_{jn}} \quad z \in Z; i \in H^z; j \in C^z; (i,j) \in P \qquad [1.39]$$

where

– h_{im} and h_{jn} are the film heat transfer coefficients for hot stream i at interval m and cold stream j at interval n, respectively;

– ΔT_{mn}^{ML} is the mean logarithmic temperature difference between intervals m and n.

In practice, the heat transfer area of a single exchanger is limited. The designer can set maximal area $A_{ij\,max}^z$ for each pair of hot stream i and cold stream j, and the existence of the corresponding heat exchanger is defined through an integer variable U_{ij}^z as follows:

$$A_{ij}^z \le A_{ij\,\text{max}}^z U_{ij}^z \qquad\qquad [1.40]$$

The objective function is to minimize the annualized total cost, which includes the utility and heat exchanger costs. The cost of a heat exchanger is linearized, including a fixed charge cost and a variable one. The total annual cost is expressed as:

$$\min \quad cost = \sum_{z\in Z}\sum_{i\in HU^z} c_i^H F_i^H \Delta T_i + \sum_{z\in Z}\sum_{j\in CU^z} c_j^C F_j^C \Delta T_j + \sum_{z\in Z}\sum_{i\in H^z}\sum_{\substack{j\in C^z \\ (i,j)\in P}} \left(c_{ij}^F U_{ij}^z + c_{ij}^A A_{ij}^z \right) \qquad [1.41]$$

where

- c_i^H and c_j^C are the hot and cold utility costs, respectively;

- ΔT_i and ΔT_j are the temperature ranges of hot stream i and cold stream j, respectively;

- c_{ij}^F is the fixed cost related to the number of shells;

- c_{ij}^A is the variable cost related to the heat exchanger surface.

Note that, when multiple matches between two streams are permitted, it is possible to add some equations and constraints to implicitly determine flow rates in splits. In this case, an additional term related to the multiple matches is added to the cost function.

1.3.1.4. Extended model

Determination of temperature intervals

The temperature partition is an essential step to guarantee the linearity of the problem. Because Barbaro's paper did not detail how to determine the temperature intervals, we propose here a solution for the temperature partition. In each zone z, the angular points of the GCC are considered to create a first set of temperature intervals. Then, the partition successively undergoes the following three steps:

– a maximal temperature step $\Delta T_{\max}^{partition}$ is set by the designer. Any higher temperature interval will be halved until the sub-intervals respect the maximal value;

– the base model requires that each stream has to own one internal interval at least. If a stream does not satisfy this condition after the first partition step, it is divided into three equal intervals;

– following the two steps above, if the total number of intervals is less than a minimum number of intervals set by the designer, the largest intervals are halved to meet the condition.

We can see that the maximal temperature step and minimum number of intervals are the key parameters for the trade-off between speed and accuracy of the algorithm. So, these parameters should fit the studied process.

Multiple HEX technologies

This section describes how different HEX technologies can be introduced in the base model. The latter will be modified, so that the algorithm will find the most appropriate technologies while satisfying the imposed constraints.

For this to happen, the variables related to heat exchangers, such as surface, number of units and cost are given an additional index t which characterizes the different HEX technologies. For this purpose, a set of technologies is defined, namely $T = \{t | t$ is an available technology$\}$.

The base model implicitly supposes that the HEX is counter-current type (equation [1.39]). In most real HEX the flow is a mixture of co-current, counter-current and cross flow. In order to take this aspect into account, a correlation factor $FHEX_t$ is introduced for each technology. This factor, set by the designer, represents the efficiency of the technology compared to the counter HEX one. In other words, the required heat transfer area $A_{1,2}$ between two flows 1 and 2 can be determined as:

$$A_{1,2}FHEX_t = \frac{q_{1,2}(h_1 + h_2)}{\Delta T^{ML} h_1 h_2} \qquad [1.42]$$

where

- $q_{1,2}$ is the heat transfer quantity;

- ΔT^{ML} is the mean logarithmic temperature difference between flows 1 and 2;

- h_1 and h_2 are the film heat transfer coefficients for flows 1 and 2, respectively.

Equation [1.39] is rewritten as follows:

$$\sum_{t\in T_{ij}} A_{ijt}^z FHEX_t = \sum_{m\in M_i^z} \sum_{\substack{n\in N_j^z ;T_n^L<T_m^U \\ j\in P_{im}^H ;i\in P_{jn}^C}} \frac{q_{im,jn}^z\left(h_{im}+h_{jn}\right)}{\Delta T_{mn}^{ML}h_{im}h_{jn}} \quad z\in Z;i\in H^z;j\in C^z;(i,j)\in P \qquad [1.43]$$

The algorithm will determine A_{ijt}^z while minimizing the cost function. A non-null surface A_{ijt}^z means that the technology t is used for heat exchange between hot stream i and cold stream j.

For each heat exchanger technology, it is possible to permit or forbid a heat exchange match between two streams i and j. For this purpose, we use a new set P_t defined as:

$P_t = \{(i,j)|$ a heat exchange match between hot stream i and cold stream j via technology t is permitted$\}$

The next equations aim to satisfy these constraints [1.44] and guarantee the consistency of calculated heat transfer areas [1.45]:

$$A_{ijt}^z = 0 \quad z\in Z;i\in H^z;j\in C^z;t\in T;(i,j)\notin P_t \qquad [1.44]$$

$$A_{ijt}^z \geq 0 \quad z\in Z;i\in H^z;j\in C^z;t\in T;(i,j)\in P_t \qquad [1.45]$$

In place of equation [1.40], the next equation will be used to determine the existence of heat exchangers matching hot stream i and cold stream j in zone z U_{ijt}^z (the maximum shell area $A_{ijt\,max}^z$ is set by the designer):

$$A_{ijt}^z \leq A_{ijt\,max}^z U_{ijt}^z \tag{1.46}$$

To take into account the cost of different technologies, the total annual cost is rewritten as:

$$\min \quad cost = \sum_{z \in Z} \sum_{i \in HU^z} c_i^H F_i^H \Delta T_i + \sum_{z \in Z} \sum_{j \in CU^z} c_j^C F_j^C \Delta T_j + \\ \sum_{z \in Z} \sum_{i \in H^z} \sum_{\substack{j \in C^z \\ (i,j) \in P}} \sum_{t \in T_{ij}} \left(c_{ijt}^F U_{ijt}^z + c_{ijt}^A A_{ijt}^z \right) \tag{1.47}$$

Flexible stream

In order to take into account the flexible streams, it is necessary for the designer to set the following sets and parameters. For a flexible stream, the inlet temperature is fixed as required for a normal stream, but the outlet temperature varies between the lower and upper values set by the designer:

$HF^z = \{i | i$ is a flexible hot stream present in zone z$\}$ ($HF^z \subset H^z$);

$CF^z = \{j | j$ is a flexible cold stream present in zone z$\}$ ($CF^z \subset C^z$);

$\left(T_{i,out}^L, T_{i,out}^U \right)$: temperature range of the outlet temperature of stream i.

The part $\left(T_{i,out}^L, T_{i,out}^U \right)$ is called the surplus part. The main idea is to consider a flexible stream as a normal stream, but we allow heat exchange between the surplus part and additional utilities, namely virtual utilities, without any cost. Specifically, a virtual hot utility is artificially created as a normal hot utility with relatively high inlet and outlet temperatures so that it can satisfy the total heat demand on every cold stream. We use the following definitions:

$$T_{i_v,out}^z = \max\left(T_m^U\big|_{m\in M^z}\right)$$

$$T_{i_v,in}^z = \max\left(T_m^U\big|_{m\in M^z}\right) + \frac{3}{2}\Delta T_{\max}^{partition}$$

[1.48]

where the index i_v refers to the virtual hot utility.

Similarly, a virtual cold utility can be defined as follows:

$$T_{j_v,out}^z = \min\left(T_m^L\big|_{m\in M^z}\right)$$

$$T_{j_v,in}^z = \min\left(T_m^L\big|_{m\in M^z}\right) - \frac{3}{2}\Delta T_{\max}^{partition}$$

[1.49]

where the index j_v refers to the virtual cold utility.

Once the temperatures are defined, the temperature partition step is carried out by using the temperature range $\left(T_{i,out}^L, T_{i,out}^U\right)$ as temperature interval bounds. This allows the following sets to be defined:

MF_i^z {m|m is a temperature interval belonging to zone z, in which the surplus zone of hot stream i is present} ($MF_i^z \subset M_i^z$);

NF_j^z {n|n is a temperature interval belonging to zone z, in which the surplus zone of cold stream j is present} ($NF_j^z \subset N_j^z$).

The following equations will be used to ensure that the virtual utilities are only allowed to match the surplus parts of the flexible streams.

For the virtual hot utility:

$$q_{i_v,m,jn}^z = 0 \quad z\in Z; j\notin CF^z; n\in N_j^z; m\in M_{i_v}^z$$

$$q_{i_v,m,jn}^z = 0 \quad z\in Z; j\in CF^z; n\in N_j^z\setminus NF_j^z; m\in M_{i_v}^z$$

[1.50]

For the virtual cold utility:

$$q^z_{im,j_v n} = 0 \quad z \in Z; i \notin HF^z; m \in M^z_i; n \in N^z_{j_v}$$
$$q^z_{im,j_v n} = 0 \quad z \in Z; i \in HF^z; m \in M^z_i \setminus MF^z_i; n \in N^z_{j_v}$$

[1.51]

Next, we define a virtual HEX technology which can be applied only to the flexible streams and virtual utilities. This can be done via the set P_i. Finally, the costs related to the virtual utilities and virtual heat exchanger technology are set to zero.

In summary, we propose a solution to consider a flexible stream whose outlet temperature is not fixed, but varies in an interval defined by the designer. The algorithm will determine the outlet temperature and the heat exchanger network which minimize the total annual cost.

1.3.2. Case study

We now present a series of results obtained via the developed model. The problem, originally presented in [SHE 95], consists of two hot streams, two cold streams, one hot utility and one cold utility. The streams data are shown in Table 1.11.

Stream	T_{in} (°C)	T_{out} (°C)	Heat load (kW)	H (W.m^{-2}.K^{-1})
H1	175	45	361	56
H2	125	65	667	56
C1	20	155	750	56
C2	40	112	300	56
H3 (utility)	180	179	–	56
C3 (utility)	15	25	–	56

Table 1.11. *Basis data of the problem (ΔTmin fixed to 20 K)*

The minimum temperature difference is assumed to be equal to 20 K for all heat exchangers used. The tested scenarios are summarized in Table 1.12. Two HEX technologies with different costs and correlation factors will be used. While HEX technology n°1 can be used for all streams, we keep n°2 only for matches between streams H1 and C1. The total costs presented are results obtained by the model. In what follows, we report the heat exchanger networks obtained. Note that splitting streams is permitted.

Test n° 1 is considered as the base test where only HEX technology n°1 is available and there is no flexible stream. Figure 1.13 shows the heat exchanger network determined by the model.

Figure 1.14 reports the heat exchanger network in test n°2 where both HEX technologies are available. We can see that this network is completely different from the one obtained from test n°1 (Figure 1.13). For the match between streams H1 and C1, HEX technology n°2 is used because it is less expensive than n°1. In addition, the heat exchanged is increased from 85 kW in test n°1 to 286 kW in test n°2, in order to get the maximum benefit from the lower cost of technology n°2. As a result, the total cost is reduced from 181 to 176 k$/year.

Common data for all tests				
		Cost ($/year)	Correction factor	Match allowed
HEX technology (A = HEX surface in m²)	n°1	5,292 + 77.8*A	1	For all matches
	n°2	4,000 + 50*A	0.7	Only for matches between streams H1 and C1
Cold utility cost	86 ($/kW/year)			
Test specifications				
	Available HEX technology	Flexible stream	Hot utility cost ($/kW/year)	Total cost (k$/year)
Test 1	Only n°1	no	173	181
Test 2	Both n°1 and n°2	no	173	176
Test 3	Both n°1 and n°2	H1 is a flexible stream with outlet temperature $\subset (45°C; 65°C)$	173	174
Test 4	Both n°1 and n°2	H1 is a flexible stream with outlet temperature $\subset (45°C; 65°C)$	1,800	598

Table 1.12. *Test conditions*

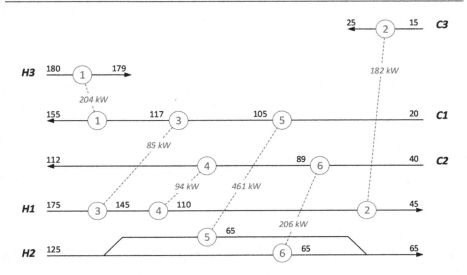

Figure 1.13. *Heat exchanger network (Test 1)*

In test n°3, both HEX technologies are available and we set stream H1 as a flexible stream with an outlet temperature $\subset (45°C; 65°C)$. Then, stream H1 consists of two parts: a principal part H1_1 where the temperature is decreased from 175 to 65°C, and a surplus part H1_2 in which the temperature varies from 65 to 45°C. Figure 1.15 shows the heat exchanger network found by the algorithm. We can observe that the surplus part H1_2 is not used. This allows the cold utility requirement to be reduced from 226 kW in test n°2 to 211 kW in test n°3. In addition, the total cost is decreased from 176 to 174 k$/year.

Using the surplus part of stream H1 would be interesting if the additional HEX cost could be counterbalanced by the drop of the hot utility requirement. For this reason, we carry out test n°4 in which the hot utility cost is set to 1,800 $/kW/year. Note that this value is very high in comparison with the hot utility costs in the real life. Indeed, it is only used to show the relevance of the flexible stream functionality proposed by our model. Figure 1.16 shows the heat exchanger network for test n°4. Hence, the total heat available on stream H1 is exploited, allowing the hot utility requirement to be reduced from 288 to 254 kW in tests n°3 and n°4, respectively.

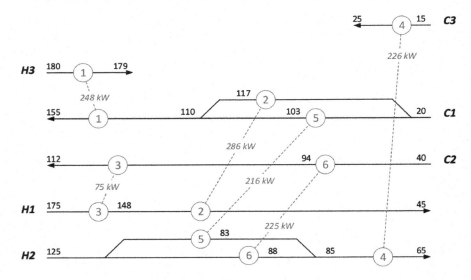

Figure 1.14. *Heat exchanger network (Test 2)*

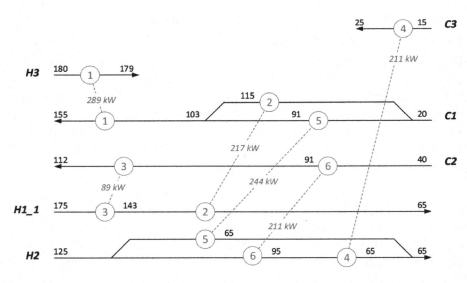

Figure 1.15. *Heat exchanger network (Test 3)*

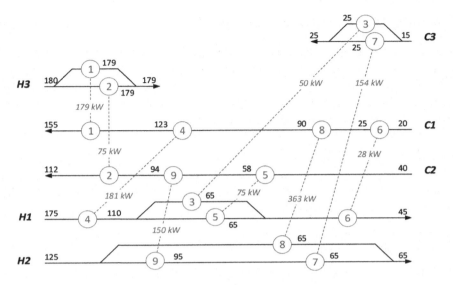

Figure 1.16. *Heat exchanger network (Test 4)*

1.4. Process modification guided by the pinch–exergy methodology

The pinch methodology enhanced with exergy concept presented earlier in this section allows us to maximize energy reuse and improves energy within a process thanks to heat recovery through the heat exchanger network (HEN) and heat conversion technologies, allowing better energy use and enhanced energy recovery. However, this methodology does not consider other parameters than heat and does not reconsider the technical choices of unitary operation.

In many cases, some technical options and operating parameters (number of stages, pressure, temperature level, etc.) in a unitary operation may be the best choice from a local point of view but could lead to missed opportunities on the process scale. Exergy analysis of the process operations can provide a picture of the operations responsible for exergy destruction; however, it does not help guide the design leading to an overall optimal solution.

1.4.1. *Process modification methodology*

Combining the process modification (based on exergy destruction criteria) and the exergy–pinch integration allows us to handle, in a systematic and comprehensive way, the process' energy efficiency improvement by acting on both energy recovery and reuse options as well as the technical choices and operating parameters of the process.

Such a combined methodology requires:

– the process to be modeled to be able to perform an exergy analysis and determine the most influent parameters. These parameters may also be identified thanks to the designer expertise:

– an energy integration using the methodology presented earlier in this section to be performed;

– this to be repeated for every technical option and possible value of the modified parameters;

– evaluate the energy/exergy consumption objective for the whole process or an economic objective, that may be the total cost, to be evaluated.

This procedure is described in Figure 1.17.

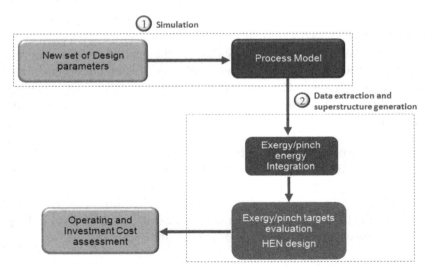

Figure 1.17. *Process modification guided by the exergy/pinch methodology*

Such a procedure may become heavy to handle manually when the number and range of operating parameters is too large. In this case, some numerical methods for metaheuristic optimization (genetic algorithms, particle swarm, etc.) may be used to handle the operating condition modification numerically until the best design is reached according to the defined objective function. This procedure is represented in Figure 1.18.

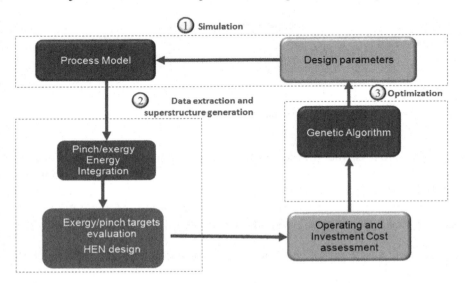

Figure 1.18. *Process modification guided by the exergy/pinch methodology and a genetic algorithm*

1.4.2. *Case study: agro food industry concentrator study*

In the agro food industry, many products or by-products need to be concentrated. This operation involves the removal of water from a solution in order to concentrate it. Many concentration techniques exist but the most commonly used techniques in the agro food industry are evaporation based.

The most basic process consists of heating the solution to a certain temperature, then it is flashed in a tank where additional heat is provided to remove the needed water content.

This process is shown in Figure 1.19.

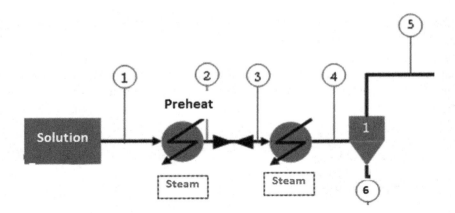

Figure 1.19. *Basic concentration process*

In this process, the solution is heated from point 1 to 2 (usually using steam). The expansion from point 2 to 3 allows the start of evaporation which is continued by heating the two-phase solution from point 3 to 4 (using steam also). It then flashes in the tank allowing the separation of the concentrated solution in point 6 and the removed water vapor in point 5.

This case study shows how the process modification methodology could be applied.

1.4.2.1. *Reference case data and pinch–exergy analysis*

In the reference process, 2,273 kg/h of the 4% starch concentrated solution arrives at 72°C and it has to be concentrated to 30%. The starch should not be heated more than 90°C for quality reasons. Therefore, it is heated in the liquid phase from 72 to 90°C before an expansion to the saturation temperature of 81°C (corresponding to an absolute pressure of 0.5 bar). The mass balance applied to the starch and water in the solution allows the mass flow rate of vapor leaving in point 5 to be determined (2,273 × (1−4%/30%) = 1,970 kg/h). In order to be able to work below atmospheric pressure, vapor at point 5 has to be condensed then pumped in a liquid state. The concentrated starch solution needs to be cooled down to 10°C before being stored.

Modeling these operations and assuming a minimum temperature difference of 10 K allows the streams presented in Table 1.13 to be extracted.

Stream	Tin (°C)	Tout (°C)	Mass flow rate (kg/h)	Heat load (kW)	Pressure (bar)
Preheating	72	90	2,273	47.5	1
evaporation	81.3	81.3	2,273	1,255	0.5
condensation	81.3	81.3	1,970	1277	0.5
Removed water cooling	81.3	30	1,970	117	0.5
Concentrated solution cooling	81.3	10	303	25	0.5

Table 1.13. *Reference concentrator streams*

Figure 1.20 shows the GCC obtained when applying the heat cascade for this process.

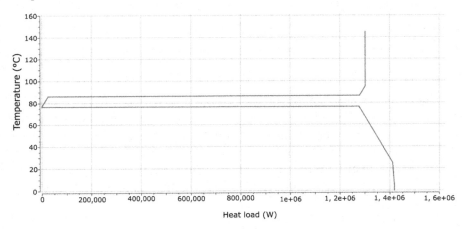

Figure 1.20. *GCC of the reference process*

The minimum heat requirement for heating is 1,302.5 kW and the one for cooling is 1,419 kW. The pinch point is located at 76.4°C (in shifted temperature).

Analyzing these results allows us to conclude first, that, no heat recovery is possible, since the heating needs are situated above the pinch and all the cooling needs are below it.

Using the methodology introduced in section 1.2 allows us to look for heat conversion systems to improve the energy use efficiency. In this case, the GCC suggests that a heat pump is adapted.

The result is shown in Figure 1.21.

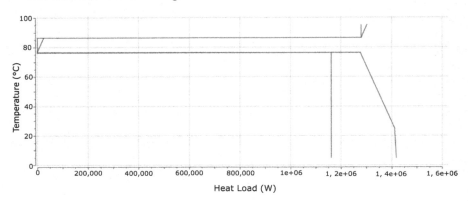

Figure 1.21. *GCC with the integration of a heat pump. For a color version of this figure, see www.iste.co.uk/zoughaib/pinch.zip*

The designed heat pump uses part of the heat released from condensing the water vapor to provide the evaporation heat. This heat pump consumes 117 kW of electricity providing 1,278 kW of heat leading to a COP of 10.9. In practice, this particular type of heat pump will use the water vapor generated by the evaporation as a working fluid and works as an open cycle by compressing this vapor and condensing it at high temperature to provide the evaporation heat.

At this step the energy use is improved, since 117 kW of electricity replaces 1,278 kW of heat coming from the steam generation utility.

However, we may think about better technical options which lead to an increase in the heat recovery potential and an further enhances the energy efficiency.

Indeed, the water vapor removed at point 5 and condensed is an important source of energy and has a high exergy potential since it is at relatively high temperature. Even if the heat pump is able to vaporize (partially) this heat, it costs an exergy consumption to operate the compressor. Allowing direct heat transfer leads to more efficient heat recovery. One possible technique is to add one more expansion step to modify the saturation temperature and allow the evaporation of part of the water by vaporizing the heat of the condensation of the first step.

1.4.2.2. *Two effects (pressure steps) and pinch–exergy analysis*

When adding the second effect, the process layout is represented in Figure 1.22.

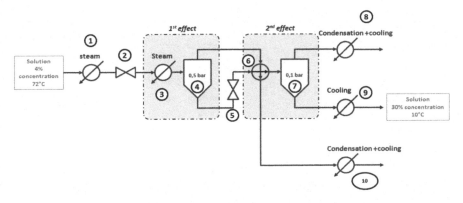

Figure 1.22. *Two-effect concentration layout. For a color version of this figure, see www.iste.co.uk/zoughaib/pinch.zip*

When adding the second effect, the expansion in step 5 is added allowing the condensation of the first-effect vapor to start in step 6 while evaporating the water of the second effect. The following steps are similar to the reference case.

This added effect introduces two degrees of freedom represented by the pressure of the second step and the quantity of water evaporated in the first step. These parameters are manipulated to globally optimize the process and its integration.

1.4.2.2.1. The first-effect evaporated water mass flow rate influence

Let us assume that the second-effect pressure is fixed at 0.1 bar (which corresponds to a saturation temperature of 46°C). This is too low to allow the heat exchange with the vapor, leaving the first effect at 81.3°C.

Let us compare two mass flow rates evaporated in the first effect and their impact on the energy integration. The first mass flow rate is 2/3 of the overall water to eliminate and the second is ½.

The two cases are modeled, streams are extracted as in the reference case and the two GCC are built and shown in Figures 1.23 and 1.24.

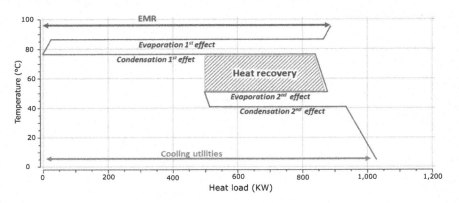

Figure 1.23. *GCC of the two-effect concentrator with 2/3 mass flow rate evaporated in the first effect. For a color version of this figure, see www.iste.co.uk/zoughaib/pinch.zip*

Figure 1.23 shows that thanks to the introduction of the second effect, a heat recovery area is now present (dashed area) that allows the second-effect water to be evaporated by condensing a part of the water vapor of the first effect. This first improvement allowed the MER to be reduced to **886 kW**. Reading the GCC suggests that the choice of 2/3 mass flow rate evaporation in the first effect is not the best choice since a large part of the condensation heat is not recovered.

Figure 1.24 shows the GCC for the case where ½ of the evaporated mass flow rate takes place in the first effect. In this case, the heat recovery area is larger, allowing most of the condensation heat to be recovered in order to evaporate the second-effect water. In this case, the MER is further reduced, reaching 677.5 kW (almost half of the reference MER).

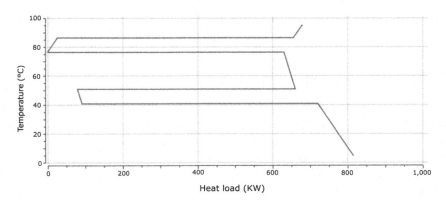

Figure 1.24. *GCC of the two-effect concentrator with 1/2 mass flow rate evaporated in the first effect*

Exploring the heat pump potential for this new design is now performed. The result is presented in Figure 1.25.

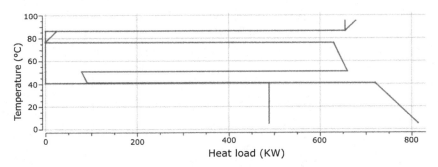

Figure 1.25. *GCC of the two-effect concentrator with the integration of a heat pump. For a color version of this figure, see www.iste.co.uk/zoughaib/pinch.zip*

Introducing a heat pump in the last design allows, as in the reference case, all the heat needed for the first-effect evaporation to be provided while recovering a part of the second-effect condensation heat. In this case, the temperature difference between the heat source and sink is much higher than in the reference case leading to a lower COP. The electricity consumption of the heat pump is 160 kW providing 650 kW of heat. The degradation of the COP is in this case greater than the gain in MER which makes this design with a heat pump less efficient from an exergy consumption point of view. The main reason for this bad performance is the COP degradation which is due to the heat source and sink temperature difference. Acting on the second-effect pressure is a way to reduce this temperature difference.

1.4.2.2.2. The second-effect pressure influence

Reducing the temperature difference between the heat sink and source of the heat pump allows the COP to be improved and hence globally reduces the exergy consumption of the process. Increasing the second-effect pressure allows this. However, this pressure increase is limited by the feasibility of the heat transfer between the first-effect vapor and the second-effect evaporator. Therefore, the highest pressure is the one that corresponds to the saturation temperature of 71.3 (0.32 bar).

Figure 1.26 shows the GCC of the two-effect concentrator with the modified pressure.

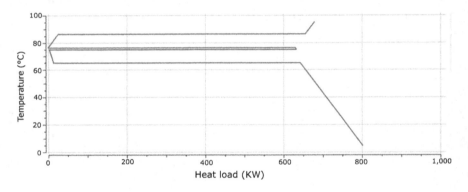

Figure 1.26. *GCC of the two-effect concentrator with the pressure modification*

Figure 1.26 shows the pressure modification allowed the temperature difference between the second-effect condensation temperature and the first-effect evaporation temperature to be reduced. The MER is identical to the case presented in Figure 1.24.

Testing the integration of a heat pump in this case is presented in Figure 1.27.

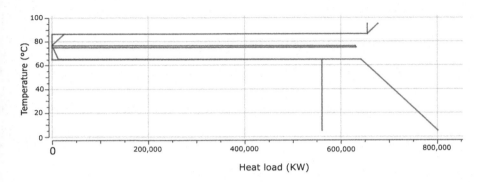

Figure 1.27. *GCC of the two-effect concentrator with the pressure modification and a heat pump. For a color version of this figure, see www.iste.co.uk/zoughaib/pinch.zip*

In this case as well the heat pump provides the needed heat for the first-effect evaporation while recovering a part of the condensation heat. In Figure 1.27, the recovered heat from the condensation is larger than that in Figure 1.25, indicating a better COP (less power for the compressor). Indeed, in this case, the compressor uses 94 kW to provide 677.5 kW. Here the COP is lower than in the reference case configuration but the MER reduction allows the compressor input power to be reduced leading to the best situation regarding the exergy consumption. Using an economic objective function may lead to a different choice since the investment costs may increase quickly in this last configuration (especially for the heat exchanger recovering the condensation heat of the first effect that presents a 10 K temperature difference).

2

Variable and Batch Processes Energy Integration Techniques: Energy Storage Optimal Design and Integration

So far, all the integration methods presented in Chapter 1 have been concerned with continuous processes. Nevertheless, other discontinuous processes exist, namely "batch" or "variable" processes.

Batch processes represent around 50% of the world chemical industrial processes in use [BIE 03]. Their implementation has increased due to their flexibility and adaptability thus allowing industrials to have an important added value to their products. These processes are frequently used in the agrochemical, pharmaceutical and chemical processing industries. The energy cost for the batch processes represents generally between 5 and 10% of the total production cost. Even if this percentage seems fair, there usually still are possibilities to reduce this cost without affecting productivity. Even more, the energy cost of the "batch processes" in some industries (dairy industries, beer production, biochemical industries, etc.) is substantial, which explains the growing interest in integrating such processes in order to reduce the cost.

Therefore, due to the importance of the discontinuous processes, it is necessary to develop methods to keep reducing their energetic cost. Some integration methods have been developed recently but there still is room for improvement.

2.1. Why is energy integration of discontinuous processes (EIDP) important?

Introducing the energetic integration into the batch processes is interesting for many reasons:

– existing batch processes rarely use the energy integration, thus we have the opportunity to retrieve lost energy;

– the study can eventually lead to other types of improvements not related to the energy cost itself, like increasing productivity or reducing production time;

– many continuous processes can employ unconsidered semi-continuous operations that can considerably reduce energy consumption;

– the EIPD methods can be used in some situations where the flow is time dependent, like in the starting and stopping phases of the continuous process or the day/night variation.

2.1.1. *Definition of batch or discontinuous processes*

In the continuous process, the flow stays continuous while traveling along the process. Its properties (temperature, mass flow, concentration, etc.) stay constant in time (except during the starting and stopping phases of the process). In the discontinuous process, conversely, the flow properties can change with time (gradually or suddenly). A flow can even totally stop while another one starts at different moments in the process. Besides, in the continuous process, the finished products are elaborated without interruption, while in the batch process, the finished product is obtained in limited quantities over a unique round of fabrication.

Figure 2.1 represents an example of a batch process [SHA 97]: products A and B are introduced into the reactor and both of them are heated to a temperature T. An exothermic reaction takes place and the temperature inside the reactor increases from T to T'. The mixture is then cooled, and sent out of the reactor before starting the cycle again.

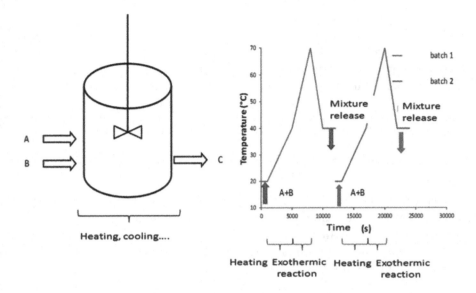

Figure 2.1. *Batch process example (chemical industry). For a color version of this figure, see www.iste.co.uk/zoughaib/pinch.zip*

The batch process is designed to accomplish one or several tasks, and it can or cannot have a defined cycle. Several flow types can be used in a single batch process [KEM 07]:

– type A: a stream operating between two fixed temperatures with the same CP, appearing and disappearing depending on the process phase;

– type B: a stream operating between two fixed temperatures but with a gradually changing CP (mass flow variation);

– type C: a stream changing temperature gradually while maintaining a constant CP (constant mass flow);

– type D: a stream changing the temperature and the mass flow (CP) gradually.

The variations in this type of processes introduce some specifications that need to be considered during the energy integration study. These specifications will be detailed in the following section.

2.1.2. *Specifications of the EIDP*

The EIDP have many specifications. Choosing between them is the main difference between the methods. Two of these specifications will be described below: the heat exchange types and the heat storage types.

2.2. Heat exchange types in discontinuous processes

There are two main heat exchange types between the operating streams in a batch process:

– direct exchange: the heat exchange is done between a cold and hot stream both in the same process phase (existing at the same time) at temperature levels which allow a direct exchange. The heat exchanger network containing the different streams allows the heat exchange, as in the energy integration of the continuous processes;

– indirect exchange: the heat exchange between the cold and hot stream is not simultaneous. Therefore, a heat storage system is needed to recover and restitute the heat at different phases of the process.

Consequently, some EIDP methods consist only of creating a heat exchange network (HEN) allowing the recovery of the energy, while other methods need a heat storage system thus increasing the energy recovery potential between different streams in the process.

2.3. Heat storage types

Many heat storage types can be used in the case of indirect exchange between streams in a given process.

2.3.1. *Sensible heat storage*

It consists of storing energy by heating a fluid or a solid without changing its phase. The fluid is then kept in a container. There are many types of sensible heat storage:

– Heat storing by mixing (Figure 2.2): the heat is stored in a way that increases the mean temperature of the container. The fluid temperature in the container is maintained homogeneous thanks to the continuous mixing.

Hence, the heat at a temperature T when stored is degraded to the mean temperature T_{av} of the container.

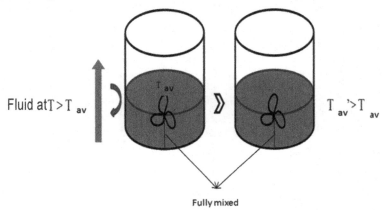

Figure 2.2. *Sensible heat storage by mixing. For a color version of this figure, see www.iste.co.uk/zoughaib/pinch.zip*

– Thermocline heat storage: this method uses the density variation of the stored fluid as a function of the temperature to obtain different fluid volumes at different temperatures in the same container by stratification. Nevertheless, and in spite of the technological advancements made in this type of storage, temperature decrease is inevitable, either by conduction or by mixing.

– Heat storage at constant temperature and variable volume: it is similar to the stratified heat storage (thermocline), but the temperature decrease is eliminated by using two different containers [STO 95]. Figure 2.3 shows the this system's operation principle. The containers have a fixed homogeneous temperature. In the storing phase (right-hand side of Figure 2.3), the fluid travels from the lower temperature container having a temperature T2 to be heated by a hot stream or a utility. When the fluid reaches the temperature T1 of the high temperature container, it is sent to be stored in that container. In the restitution phase (left-hand side of Figure 2.3), the hot fluid from the high temperature container transfers the heat to a cold stream. When the fluid reaches T2, it continues its way to the low temperature container.

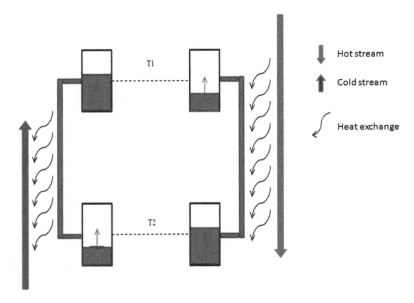

Figure 2.3. *Storage principle at "constant temperature, variable volume".*
For a color version of this figure, see www.iste.co.uk/zoughaib/pinch.zip

2.3.2. *Latent heat storage*

This type of storage uses phase changing materials to store/restitute heat
by using a solidification ⇔ fusion process. The phase changing temperature
depends on the used material. Generally, this method allows us to reduce the
storage volume used compared with the sensible heat storage method.

2.4. Different energy integration methods for discontinuous processes

After presenting EIDP specifications (different thermal exchange types
(direct and indirect) between streams in a batch process as well as the
different heat storage methods), the different EIDP methods will be
presented in this section. Similarly to the energy integration in continuous
processes, many integration methods can be used in discontinuous processes.

Some adapted from pinch analysis, presented previously in Chapter 1, for the batch processes. Others use mathematical formulations to build specific energy integration designs for discontinuous processes.

2.4.1. *Pinch analysis-based methods*

Even though this method was first designed to energetically integrate continuous processes, it can be adapted, after modifications [KEM 07], for batch processes. Many of the previous works presented adaptations of the pinch analysis method to the EIDP, thus presenting different evaluation methods for energy recovery and architecture design for energy integration in discontinuous processes.

2.5. Time average model

The pioneering works presenting the adaptation of pinch analysis to discontinuous processes [CLA 86a, CLA 86b], used the "Time Average Model" (TAM), in order to evaluate the maximum energy objective. In this model, all streams in a given cycle of the batch process are integrated in time. Consequently, the composite curves and energy cascades, used normally for streams in the continuous process, are also used for the integrated streams in the discontinuous process. Thus, the composite curves, (heat (kJ)-Temperature (°C)) or (heat (kWh)-Temperature (°C)), can be drawn. These curves can then be used to calculate, for cooling and heating, the minimum needed energy for the process. This method helps calculate the hypothetical minimum energy needs which are not feasible since the conditions for the direct heat exchange are not satisfied.

To illustrate this model, let us consider the example presented in Table 2.1. The streams, their heat capacity flux along with their beginning and ending times, are presented. The heat of the different streams is given in kJ to apply the TAM method. Then the pinch analysis is used on the time-integrated streams. In practice, the CP (kW/K) used in the pinch analysis method are replaced with CPxΔt (kJ/K) in the TAM (where Δt is the time slot where the stream exists). The same methodology as in the pinch analysis is then used to draw the composite curves (heat (MJ)-Temperature (°C)) of the hot and cold streams as shown in Figure 2.4, considering $\Delta T_{min} = 10$ K.

Stream number	Stream type	T_e (°C)	T_s (°C)	CP (kW/K)	t_e (hours)	t_s (hours)	CP.Δt (kJ/K)	Heat (MJ)
1	Cold	90	170	10	0	0,5	18,000	1,440
2	Hot	120	20	4	0	0,5	7,200	720
3	Cold	40	160	8	0,5	1	14,400	1,728
4	Hot	135	20	6	0,5	1	10,800	1,242

Table 2.1. *Example of the stream in a batch process*

The heat recovery by heat exchange between hot and cold streams using this model is 1,242 MJ. The minimum energy requirements are 1,926 MJ for hot utility and 720 MJ for cold utility. The pinch point is at 95°C. However, these are in fact hypothetical values as mentioned earlier. In fact, the streams presented in Table 2.1 are not existent at the same time, thus the recoverable heat quantity cannot be obtained by a simple direct exchange between simultaneous streams.

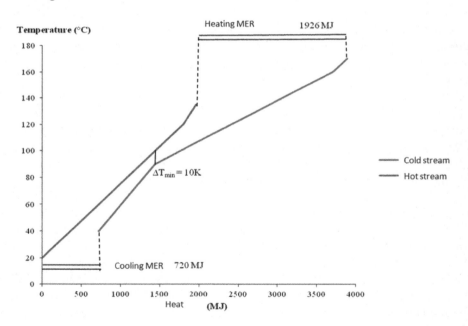

Figure 2.4. *Composite curves related to the TAM. For a color version of this figure, see www.iste.co.uk/zoughaib/pinch.zip*

2.6. Time slice model

Since the operating streams in the batch process rarely exist at the same time, the TAM gives the theoretical energy target but it cannot be used to calculate feasible energy targets. Therefore, some previous works [KEM 89a] presented the "Time Slice Model" (TSM), in which the process is divided into time periods. The beginning of a period and the end of the previous one are determined whether one of these events occur:

– the stream operating temperature changes;

– the specific heat capacity or the mass flow of an operating stream changes;

– a new stream appears or one disappears.

As a result, this method can only be used on the streams of type A and on a time period, long enough so the streams can be considered in a permanent regime in a given period. It can also be used with other stream types if the mean value of the variable is considered in each interval. This in fact is an approximation that can be very optimistic in some cases.

Thereafter, the classic pinch analysis method is used on all the intervals which allow us to draw the energy cascades and the composite curves of each interval. Next the energy recovery potential can be obtained by interval. These energy objectives are realistic and can be achieved. The heat exchange network, namely "Maximum Heat eXchanger network" (MXH) can now be designed [KEM 89b]. This network is made of exchangers that allow a direct exchange between different streams in the same time interval. We should also mention that such network can contain heat exchanges crossing the global pinch point (the global pinch point can be obtained from the TAM). For that reason, energy objectives calculated using the TAM cannot be obtained via the TSM (it would be a violation of one of the pinch method rules).

The TSM will now be applied to the example presented in Table 2.1 to evaluate the heat quantity that can be recovered for this process by performing a direct exchange between existing streams in one time interval. The time is here divided into two intervals containing each a hot and a cold stream. Afterwards, the modified pinch analysis method is applied in each interval to obtain the composite curves for the streams for the first time interval in Figure 2.5(a) and for the second time interval in Figure 2.5(b).

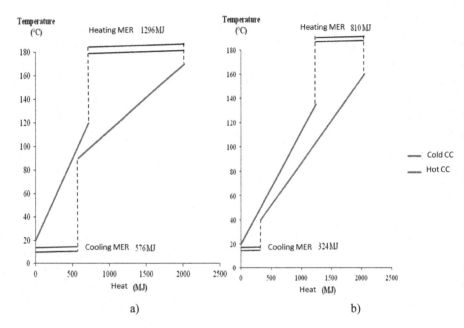

Figure 2.5. *Composite curves related to the TSM in the first (a) and second (b) time intervals. For a color version of this figure, see www.iste.co.uk/zoughaib/pinch.zip*

In the first time interval, the pinch point is at 95°C and the maximum recoverable heat between the two streams is 144 MJ. In the second time interval, the pinch point is at 45°C and the maximum recoverable heat is 918 MJ. Hence, the recovered energy by direct exchange is 1,062 MJ, when the value obtained using the TAM was 1,242 MJ.

2.7. Rescheduling

The TSM applied to the process streams allows non-negligible energy recovery (even if it is far from the theoretical potential). However, the economic viability of the MXH can be questionable since the investment is important, while the use of most of the heat exchangers is part of the time leading to less operating costs compared to the investment. In addition, in many batch processes few or some streams exist simultaneously which means that the TSM is useless.

Alternatively, rescheduling the process operations [KEM 90] may help increase the heat recovery using the TSM. When rescheduling the process operations, the streams are not modified (temperature and mass flow rate); however, their starting time may be adapted to maximize the simultaneous occurrence time.

We can list many rescheduling types:

– when several batch operations are performed in series and in parallel. In a series, the starting time of a batch may be modified to take advantage of some streams that may exchange heat with another batch operation (creating a cycle phase difference between parallel cycles);

– in a batch process, the start, the end and/or the duration of operations may be modified in a way to allow the simultaneous existence of streams that can exchange heat.

Before proceeding with rescheduling, the Gantt diagram (time–event) is plotted allowing to visualize the streams occurrence with time. Based on this diagram, many scheduling modifications can be studied and analyzed based on their feasibility (in terms of the process organization) and in terms of the increase of the heat recovery potential.

The Gantt diagram of the streams presented in Table 2.1 is shown in Figure 2.6.

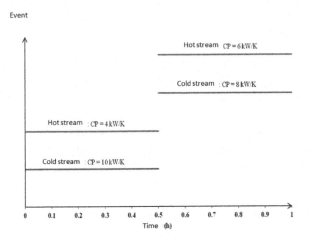

Figure 2.6. *Gantt diagram (time–event). For a color version of this figure, see www.iste.co.uk/zoughaib/pinch.zip*

In practice, for each scheduling proposal a TSM is performed until the best configuration is found.

Even if this approach increases consequently the heat recovery potential; it also presents many applicability issues in existing sites. Indeed, it is not always possible to modify the production sequence or to introduce parallel operations. Finally, the flexibility of the rescheduled process is reduced.

2.8. Introducing heat storage

Also based on the TSM, an alternative to rescheduling, heat storage may be imagined to transport energy from a time interval to another instead of changing the process operations' schedule.

Using the TSM, the process may be analyzed for each time interval thus plotting the composite curves, the Grand Composite Curve (GCC) and calculating the heat cascade.

For each time interval, the GCC allows us to identify the period pinch point, the surplus heat below the pinch and also the deficit heat above the pinch. Based on the model proposed by [DIN 02] and using the pinch methodology rules presented in Chapter 1, a surplus heat available in a certain time interval will be useful in an another time interval only if its temperature is higher than the pinch point of the second interval. If this amount of heat is stored, it allows us to increase the recovery potential determined using the TSM.

In practice, when comparing the GCC's of the process for each time interval, it is possible to identify the heat respecting the above condition (e.g. available between the pinch points of both time intervals). This heat is potentially storable which will help to increase the heat recovery potential. To design such a heat storage tank, we have to design a virtual cold utility that recovers the heat while matching (in energy) a hot utility at the same temperature for another time period.

When considering the example of Table 2.1, we can note that the pinch point of the first time period is higher than that in the second time period. Therefore, as shown in Figure 2.7, the energy available in the temperature interval between the first time period pinch point and the second time period pinch point is partially recoverable by storing it.

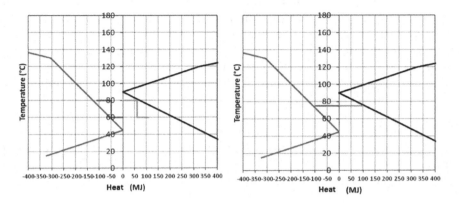

Figure 2.7. *Heat storage integration. For a color version of this figure, see www.iste.co.uk/zoughaib/pinch.zip*

In order to reach the target defined by the TAM, an infinite number of heat storage tanks should be used to recover 180 MJ from the first time period (black GCC) which corresponds to the deficit of the second time period (light gray GCC) between 95 and 45°C (both periods pinch points).

In practice, a limited number of heat storage tanks are used. Once the number of heat storage tanks are defined, the design task is to define, as described before, the virtual hot and cold utilities matching the excess or the surplus at the same temperature.

As an example, in Figure 2.7, we show on the right the energy recovered and stored in one heat storage tank and on the left the case where we have two heat storage tanks.

On the right-hand side of Figure 2.7, we can see that the heat storage tank's temperature is 75°C, which both the heat surplus (from the first time period) and the heat deficit (of the second time period) match. The recovered and stored heat is 100 MJ (more than half of the heat deficit of the second time period which was between 95 and 45°C).

When using two heat storage tanks, the aim is to choose the two temperatures that maximize the amount of recovered heat. As we can see on the left-hand side of Figure 2.7, recovering all the heat in surplus at 80°C in the first time period (60 MJ) is completed by the whole heat deficit at 60°C (50 MJ) which is largely available from the first period. Finding the temperature match of 60°C is graphically possible by plotting a vertical line from the end of the virtual utility line at 80°C. Its intersection with the GCC gives the matching temperature. Adding a second storage tank allows to therefore increase the heat recovery by 10% in this case.

This methodology allows us to graphically determine heat storage tank capacities and temperatures that will allow us to reach better energy recovery targets than with the simple TSM and this also thanks to the energy storage keeping higher flexibility for the process.

However, with a high number of periods and a more complex process (with a high number of streams), the combinatory becomes so high that the graphical method finds its limit. Adding the heat conversion systems to the heat recovery architecture further increases the complexity; therefore, mathematical programming is a better solution to determine a systematic and rigorous integration solution.

2.8.1. *Exergy-based model for heat integration of variable processes designing heat storage tanks and heat conversion systems*

The proposed model to address the problem of heat integration in batch or variable processes is formulated as a Mixed Integer Linear Programming (MILP) optimization problem. The model optimizes a linear objective function based on the consumed exergy in a similar manner as the model presented in Chapter 1. This model is developed by Salame *et al.* and published in [SAL 16] and [SAL 15].

The heat integration problem in batch plants consists of finding a heat exchangers' network that is capable of energetically integrating the variable streams in time. Knowing that the streams are variable in time, as shown earlier, thermal storage is the solution to obtain maximum heat recuperation over the whole cycle. The use of heat conversion systems allows us to improve further the heat recovery while reducing exergy consumption (see Chapter 1).

2.9. Description of heat integration network

Since heat storage tanks are used, it is not feasible to perform direct heat exchange. Therefore, an intermediate fluid circulates in the network and its role is to transport heat from the hot streams to the cold streams or from/to the thermal storage tanks (Figure 2.8).

At this point, the integration problem is divided into sub problems depending on the type of thermal storage tank used. Here, we will treat one type of thermal storage tank: the case where the fluid circulating in the network is stored in tanks at a constant temperature and variable volume [STO 95].

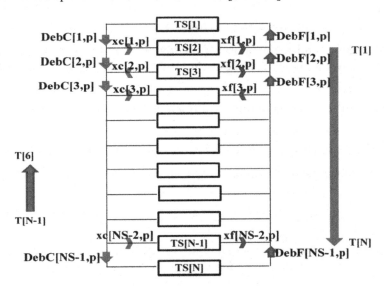

Figure 2.8. *Heat integration network including thermal storage tanks of the fluid circulating in the network. For a color version of this figure, see www.iste.co.uk/zoughaib/pinch.zip*

2.10. Inputs and outputs of the model

The inputs of the model are the following:

– the number of streams to heat "NFF" and those to cool "NFC";

– calculation time is discretized to a certain number of periods "NP", each period's duration is "tp";

– the discretized temperature range parameters (maximum temperature, minimum temperature and node temperatures); this allows us to linearize the model;

– the minimum temperature difference of the heat exchangers between the fluxes and the circulating fluid in the network "Pinc";

– the reference temperature, as well as the temperature of the hot source and cold source for exergy calculations;

– the maximum number of desired heat storage tanks "nb_Stocks";

– the maximum number and type of heat conversion systems.

The outputs of the model are the following:

– the number and temperature of thermal storage tanks used, in addition to their maximal capacity and the variation of the heat stored as a function of time;

– the number and operating conditions of each heat conversion systems used;

– the heat exchanged between the different streams in each time period;

– the net heating and cooling utilities needed in each time period.

2.11. MILP formulation of the optimization problem

Each storage tank is filled or emptied depending on the heat or cooling demands of the streams. Hence the volume of the fluid stored in any thermal storage tank varies in time leading to a corresponding mass flow in each branch of the network.

Many parameters are precalculated in the algorithm including, the theoretical number of storage (NS) tanks and their temperatures (equation [2.1]) (Figure 2.9):

$$TS[k] = T\left[floor\left(\frac{k+1}{2} \right) \right] + if\,(i\ mod\,2 = 0)\ then\,(-Pinc)\ else\,(+Pinc) \qquad [2.1]$$

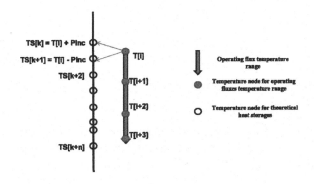

Figure 2.9. *Storage tanks theoretical temperatures. For a color version of this figure, see www.iste.co.uk/zoughaib/pinch.zip*

where TS[k] is the temperature of the heat storage tank k. T[i] is the temperature of the node i of the shifted temperature range. Pinc is the minimum temperature difference of the exchanger between the intermediate fluid circulating in the network and the operating fluxes of the process.

These temperatures are then arranged from the highest to the lowest temperature.

In addition to the storage tank temperatures, the coefficients of performance (COP), the efficiencies of the heat conversion systems are precalculated for each couple of storage tank temperature (for HP and absorption chillers) and for each storage tank temperature (for CHP and ORC). The same reduced order model based on an exergy efficiency (second law efficiency) and an ideal cycle COP or efficiency is used. The formulation is identical to the one presented in equations [1.4]–[1.8].

The constraints on the variables are the following:

– Energy balance applied to each interval of temperature "i" of each hot stream "j" at a certain time period "p": heat that exists in the stream will be either rejected to a cold utility or exchanged with the fluid circulating in the network between two thermal storage tanks at lower temperatures.

$$CP_c[i,j,p] \times (T[i] - T[i+1]) = UF[i,j,p] +$$

$$\left(if \ T[i] \geq (TS[k] + Pinc) \ and \ T[i+1] \geq (TS[k+1] + Pinc) \right) \qquad [2.2]$$

$$\sum_{k=1}^{NS-1} DebF\_d[k,i,j,p] \times (TS[k] - TS[k+1])$$

where UF is the cooling power and DebF_d is the partial heat capacity flow (m.Cp) that exchanges with the hot stream (the Cp of the intermediate fluid is assumed constant). This is represented in Figure 2.10.

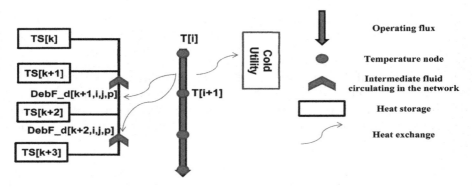

Figure 2.10. *Energy balance applied to an interval of temperature i of a hot flux j at a certain period of time. For a color version of this figure, see www.iste.co.uk/zoughaib/pinch.zip*

– Energy balance applied to each interval of temperature "i" of each cold stream "j" at a certain period of time "p": the heat needed is either supplied by a hot utility or by the fluid circulating in the network between two thermal storage tanks at higher temperatures.

$$CP_f[i,j,p] \times (T[i] - T[i+1]) = UC[i,j,p] +$$

$$\left(if \ TS[k] \geq (T[i] + Pinc) \ and \ TS[k+1] \geq (T[i+1] + Pinc) \right) \qquad [2.3]$$

$$\sum_{k=1}^{NS-1} DebC\_d[k,i,j,p] \times (TS[k] - TS[k+1])$$

where UC is the heating power and DebC_d is the partial heat capacity flow that exchanges with the cold stream.

– Energy balance applied to each thermal storage tank "k" at a certain period of time "p": the storage tank heat capacity $Stock[k,p]$ is calculated in kJ/K. The heat power stored in a thermal storage tank "k" at the period of time "p" is equal to the difference of circulating fluid heat capacity flows entering and exiting this storage tank.

$$Stock[k, p+1] = Stock[k, p] + tp[p] \times$$
$$(-xc[k-1, p] - xf[k-1, p] +$$
$$\frac{UCS[k, p]}{TS[k] - TS[k+1]} - \frac{UFS[k, p]}{TS[k] - TS[k+1]} -$$
$$\frac{UCS[k-1, p]}{TS[k-1] - TS[k]} + \frac{UFS[k-1, p]}{TS[k-1] - TS[k]}$$

[2.4]

where $UCS[k,p]$ and $UFS[k,p]$ are respectively the heating and cooling powers of the utilities between two storage tank temperatures $TS[k]$ and $TS[k+1]$.

$P_{cond,HP}$ and $P_{evap,HP}$ are the heat loads at the condensers and evaporators of heat pumps. $P_{evap,ORC}$ is the heat absorbed by the ORC boiler.

$xc[k-1,p]$ is the heat capacity flow entering the storage tank "k" after having exchanged with the cold stream or exiting the storage tank to exchange heat with the cold stream. $xf[k-1,p]$ is the heat capacity flow entering the storage tank "k" after having exchanged heat with the hot stream or exiting the storage tank to exchange heat with the hot stream.

– Node law applied to all the heat capacity flows entering or exiting each knot "k" in the circuit at a certain period of time "p" (Figure 2.11):

$$DebF[k+1, p] - DebF[k, p] + xf[k, p] = 0$$

[2.5]

$$DebC[k, p] - DebC[k+1, p] + xc[k, p] = 0$$

[2.6]

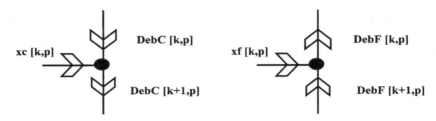

Figure 2.11. *Node law applied to each node in the network (law of mass conservation)*

where xf and xc have positive values if the flow is entering the knot and negative values if it is leaving it.

– Energy balance applied to interval "k" between two thermal storage tanks TS[k] and TS[k+1]:

$$\text{DebF}[k,p] = \sum_{i=1}^{NT-1} \sum_{j=1}^{NFC} \text{DebF\_d}[k,i,j,p] \qquad [2.7]$$

$$\text{DebC}[k,p] = \sum_{i=1}^{NT-1} \sum_{j=1}^{NFF} \text{DebC\_d}[k,i,j,p] \qquad [2.8]$$

where Deb_d and DebC_d are the partial heat capacity flows. *DebF* and *DebC* are the total heat capacity flows in the network.

– A constraint that imposes the equality between the initial state of a thermal storage tank "k" and its final state:

$$Stock[k,1] = Stock[k, NP+1] \qquad [2.9]$$

– A constraint that forces every storage tank "k" to be emptied at least for one period of time:

$$\sum_{p=1}^{p=NP+1} \text{bS\_n}[k,p] \le NP \qquad [2.10]$$

where $bS\_n[k,p]$ is a binary that is equal to 1 if the storage tank k has a positive value at a period "p" and 0 if this storage tank is empty (i.e. value = 0). Knowing that there are NP periods of time, each storage tank k should have NP+1 values, one of them is imposed to be 0.

– A constraint that forces the number of storage tank to be under a certain parameter desired by the user:

$$\sum_{k=1}^{k=NS} bS[k] \le nb\_Stocks \qquad [2.11]$$

where bS is a binary that is equal to 1 if the storage tank does exist and is equal to 0 if it does not. nb_Stocks is the maximum number of storage tanks desired by the user.

2.12. Function to optimize

The objective of heat integration is to reduce net heating and cooling demands in exergy. Hence, the function to minimize is the sum of the hot, cold and heat conversion system exergy consumption in the process:

$$Minimize : \sum \left(Ex_c + Ex_f + ExS_c + ExS_f + Ex_{HP} - Ex_{ORC} \right) \qquad [2.12]$$

where:

$$Ex_c = \sum_{i=1}^{NT-1} \sum_{j=1}^{NFF} \sum_{p=1}^{NP} UC[i,j,p] * tp[p] * \left(1 - \frac{T_{ref}}{T_{SC}} \right) \qquad [2.13]$$

$$ExS_c = \sum_{i=1}^{NT-1} \sum_{j=1}^{NS} \sum_{p=1}^{NP} UCS[i,j,p] * tp[p] * \left(1 - \frac{T_{ref}}{T_{SC}} \right) \qquad [2.14]$$

Ex_c and ExS_c are the exergies required for heating.

The exergies consumed to satisfy the cooling needs, using a cold utility at a certain temperature T_{SF} which is lower than reference temperature T_{ref}, are calculated in equations [2.15] and [2.16]:

$$Ex_f = \sum_{i=1}^{NT-1}\sum_{j=1}^{NFC}\sum_{p=1}^{NP} UF[i,j,p] * tp[p] * \left(1 - \frac{T_{SF}}{T_{ref}}\right) \qquad [2.15]$$

$$ExS_f = \sum_{i=1}^{NT-1}\sum_{j=1}^{NS}\sum_{p=1}^{NP} UFS[i,j,p] * tp[p] * \left(1 - \frac{T_{SF}}{T_{ref}}\right) \qquad [2.16]$$

And finally, the exergy consumed by the HP, which is the power of the compressor, is calculated as well as the exergy valorized by the ORC, which is the power generated by the turbine:

$$Ex_{HP} = \sum_{i=1}^{N_s-2}\sum_{j=i+1}^{N_s-1}\sum_{mp=1}^{N_{mp}}\sum_{Mp=1}^{N_{Mp}} P_{comp,HP}[i,j,mp,Mp] \times t_{mp}[mp] \qquad [2.17]$$

$$Ex_{ORC} = \sum_{i=1}^{N_s-2}\sum_{j=i+1}^{N_s}\sum_{mp=1}^{N_{mp}}\sum_{Mp=1}^{N_{Mp}} P_{turb,ORC}[i,j,mp,Mp] \times t_{mp}[mp] \qquad [2.18]$$

2.13. Case studies

2.13.1. *Molecular sieve regeneration process*

A molecular sieve is a material with very small holes of precise and uniform size. These holes are small enough to block large molecules, while allowing small molecules to pass. Many molecular sieves are used as desiccants.

Molecular sieves are used in different types of applications. They are used for the purification of gas streams in the petroleum industry. For example, in the liquid natural gas industry, the water content of the gas must be reduced to very low values to prevent it from freezing (which causes blockages) in the cold section of liquid natural gas plants.

Molecular sieves are also used in chemistry applications for compound separation and drying reaction starting materials. They are also used in the filtration of air supplies for breathing apparatus.

Molecular sieves used for dehydration need to be dried before being used again. This drying process is consumes energy. This case study treats the possibility of heat integration including heat storage tanks in the drying process of a molecular sieve.

This drying process is done in two steps:

– first, hot dry air is blown into the molecular sieve. It leaves the molecular sieve after being charged with water vapor. The molecular sieve is then left hot and dry;

– then, dry air at lower temperature is blown to cool the molecular sieve.

This molecular sieve is now ready to be used again in dehumidification processes.

2.13.2. *Data extraction and TAM application*

An example of a drying process of a molecular sieve is illustrated in Figure 2.12. This figure represents the variation of air temperatures Tin, blown into the molecular sieve and Tout, after having left the molecular sieve. The process, as shown in the figure, is divided in two phases. In the first phase, dry air at 290°C is blown to the molecular sieve and leaves charged with humidity at a lower temperature Tout. The second phase is the cooling phase where the air is blown at 45°C and leaves at a higher temperature Tout.

The heating and cooling utilities needed in this application are the following:

– heating the dry air from a temperature of 45–290°C to be blown into the molecular sieve in the first phase;

– cooling the rejected humid air resulting from the drying process in the first phase from Tout to 45°C;

– cooling the rejected air resulting from the cooling process from Tout to 45°C in the second phase (the cooling phase).

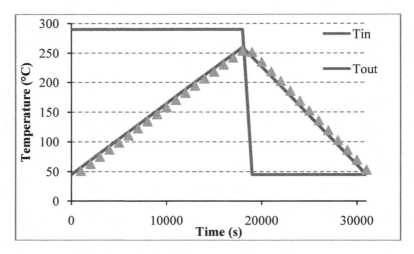

Figure 2.12. *Variation in time of the air temperature
used for molecular sieve drying. For a color version
of this figure, see www.iste.co.uk/zoughaib/pinch.zip*

To study the heat integration possibilities by using a multi-period approach, the total time of the cycle is divided into equal time intervals; in each interval the inlet and outlet temperatures of the air are given. To be more accurate, the output temperature in each period is calculated as the average of the output temperature (equation [2.20]) on the whole period between two instants t and t + dt.

$$Tout\_av = \frac{Tout_{(t)} + Tout_{(t+dt)}}{2} \qquad\qquad [2.19]$$

A first estimation of the maximum heat integration potential in this case is done by integrating all the streams over the whole cycle. The TAM is then applied and CC in the energy term is illustrated in Figure 2.13.

This analysis shows that for a 10 K pinch minimum temperature difference, the minimum hot and cold utilities required after all the possible exchanges made between all the streams are 2,792 MJ for cooling and 6,257 MJ for heating. Heat exchanged between the different fluxes is almost 7,924 MJ.

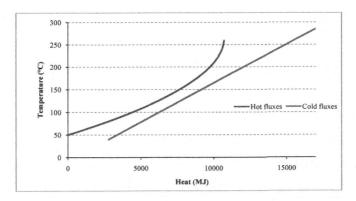

Figure 2.13. *Pinch analysis (TAM) applied to the integrated problem of drying of a molecular sieve. For a color version of this figure, see www.iste.co.uk/zoughaib/pinch.zip*

It should be noted, as explained earlier, that the integration, which assumes that the streams can be used in a heat exchanger at any period, is not feasible in practice but only constitutes a theoretical target. Hence, heat storage tanks have to be used to allow the heat exchange between non-simultaneous fluxes and thereby to try and approach the theoretical target. The methodology presented in section 2.4 allows us to propose a practical design of the heat storage tank system. More precisely, it helps to determine the temperature, capacity and variation of the energy stored in time.

2.13.3. *Heat storage tank design*

To study the heat integration possibilities four scenarios are proposed, each with a different maximum number of heat storage tanks. The algorithm detailed in section 2.4 will be used and results are compared.

It should be noted that the following parameters were given to the algorithm:

– the number of periods is 31 of 1,000 s each;

– the number of streams to heat is 18 and of those to cool is 31;

– the heat capacity of all the flows is equal to 3.216 kW/K;

– the minimum temperature difference of the heat exchangers between the fluid circulating in the network and the operating fluxes is considered to be 5 K (which allow us to eliminate the penalty of intermediate fluid since in the TAM it was assumed to be 10 K);

– for exergy calculations: the reference temperature is equal to the cold source temperature which is equal to 15°C. The hot source temperature is considered to be 900°C;

– the maximum number of heat storage tanks is a parameter that will be changed from one simulation to another.

Many scenarios are simulated. In each simulation, the maximum number of heat storage tanks is varied. Then the exergy consumed and destroyed is calculated for each solution and is illustrated in Figure 2.14. It should be noted that if no heat integration is done, the total exergy needed for this application is almost 10,820 MJ. This value is not represented in Figure 2.14.

The continuous line in the graph represents the minimum theoretical required exergy of the integrated problem (TAM); this is the target value of the exergy. The results show that just by proceeding to a heat exchange between the hot fluxes and cold fluxes in the first 18 periods of time (TSM), there is a saving potential of 3,984 MJ of exergy (this case is represented on the graph by the point at zero storage tanks). In other terms, if only heat exchangers are installed between the simultaneous streams, 36.8% of the total spent exergy is saved.

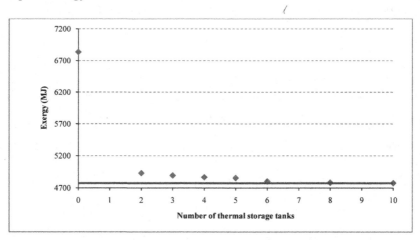

Figure 2.14. *Exergy required to accomplish the integrated drying process as a function of the number of thermal storage tanks used*

Using two heat storage tanks reduces the spent exergy by 54.4%. This is represented on the graph by the point at two heat storage tanks.

As it is noticeable also on the graph, the values of exergy consumption for the cases where the number of storage tanks is above three are very close. This means that increasing the number of thermal storage tanks does not have a major effect on reducing the exergy demands of the process.

To understand the type of solutions given by the algorithm, the case with two maximum heat storage tanks will be detailed and discussed.

Assuming that under pressure water is the intermediate fluid circulating in the network, Figure 2.15 shows the variation of the volume of water stored in both storage tanks in one cycle. The storage tank temperatures are 240.2 and 56°C.

It should be noted that the process is cyclic, in other words it is repeatedly done over the whole production time.

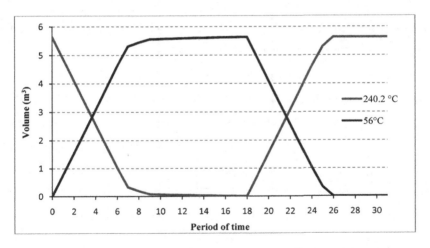

Figure 2.15. *Volume variation within time in m³ of the thermal storage tanks used in the heat integration of the drying process. For a color version of this figure, see www.iste.co.uk/zoughaib/pinch.zip*

In the first 18 periods of time (first phase), there are simultaneous cooling and heating demands. In this phase, the behavior of the thermal storage tanks change between the beginning and the end of the phase. We can see on the graph that the high temperature storage tank (240.2°C) is emptied between the first and tenth period of time. The low temperature storage tank (56°C) is simultaneously filled at the same rate. Afterwards, the volumes of these storage tanks remain almost constant. A reversed behavior is observed in the second phase (from period 18 to 31). In this phase, the high temperature storage tank is filled from period 18 to period 26 and the low temperature storage tank is filled at the same rate. Afterwards, the volumes of both storage tanks remain constant.

Due to this change in storage tanks' behavior in both of the phases, each phase is divided into two sub-phases. The results of the simulations also allow us to analyze the behavior of the network in each phase, more specifically in each sub-phase. This allows us to determine the architecture of the heat integration network all over the cycle. Figure 2.16 shows the topology of the heat exchangers' network.

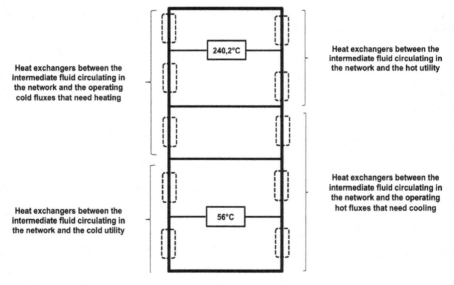

Figure 2.16. *The heat integration network scheme in the drying process of a molecular sieve*

As previously said, the operating scenarios of this heat integration network vary from one sub-phase to another. Figure 2.17 represents the heat integration network operation in the sub-phases of the first phase.

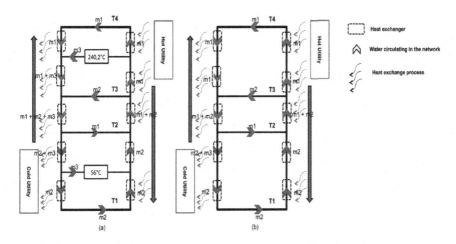

Figure 2.17. *Heat integration scenario in the first phase of the drying process.*
For a color version of this figure, see www.iste.co.uk/zoughaib/pinch.zip

In this phase, both cooling and heating demands exist and it is divided as already mentioned into two sub-phases represented in Figure 2.17(a) and (b), respectively. In the first sub-phase (Figure 2.17(a)), a water mass flow rate m3 leaves the high temperature storage tank (240.2°C) and exchanges its heat with the cold streams of the process. The rest of the heat (if it exists) is exchanged with a cold utility, and the water mass flow rate m3 is stored in the low temperature storage tank (56°C). This flow rate has to transport the heat already stored from the previous cycle at high temperature to the cold streams, and in doing so responds to a large part of the heating demands of the process. At the same time, a mass flow rate m2 is heated from the lowest temperature in the network T1 to a certain temperature T3 by exchanging heat with the hot streams of the process. This mass flow rate exchanges then a part of its heat with the cold streams and then it is cooled using the cold utility to temperature T1 to continue its circulation in the network. This sub-network's main function is to ensure the cooling demands of the hot streams of the process. Simultaneously, a mass flow rate m1 is heated from a certain temperature T2 to T3 by heat exchange with the hot streams, and then it is heated using a hot utility to the highest temperature in the network T4. Then its heat is exchanged with the cold streams which cools it to the temperature T2 to continue its circulation in the network. This sub-network's main function is to ensure the heating demands of the cold streams of the process.

In the second sub-phase (Figure 2.17(b)), the volumes of thermal storage tanks are constant which means the mass flow rate m3 exchanged between them does not exist anymore. Two mass flow rates continue to circulate in the two sub-networks previously described.

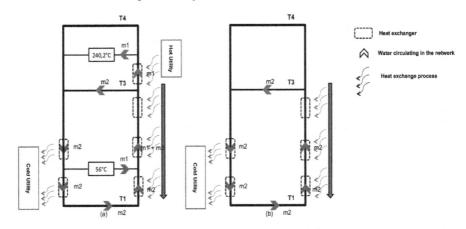

Figure 2.18. *Heat integration scenario in the second phase of the drying process. For a color version of this figure, see www.iste.co.uk/zoughaib/pinch.zip*

In the second phase (Figure 2.18), only cooling demands exist. In the first sub-phase, a water mass flow rate m1 leaves the low temperature storage tank to be heated to a certain temperature T3 by exchanging heat with the hot streams of the process. Heating up to the temperature of the storage tank (240.2°C) is done using a hot utility. The mass flow rate m1 is then stored in the high temperature storage tank. This sub-network has to store the heat available in this phase to be used in the next cycle. At the same time, a certain water mass flow rate circulates between the lowest temperature of the network T1 and temperature T3 by exchanging heat with the streams to cool. This heat is then exchanged with a cold utility and the flow continues to circulate in the sub-network. This sub-network's function is to exchange the available heat that is not stored in this phase with the cold utility. In the second sub-phase (Figure 2.18(b)), the storage tanks' volumes are constant, no heat is stored and then only the mass flow rate m2 continues to circulate in the sub-network.

2.14. The dairy cleaning processes

The case study, presented in [MUR 11], is the heat integration possibility in the dairy industry where waste water mainly comes from cleaning and concentration processes. The process uses hot water and heated cleaning solutions, where heat is provided by a hot utility. The demands of the process are shown in Table 2.2.

Demands	Parameters				
	Tin(°C)	Tout(°C)	M(kg/s)	P(kW) (instantaneous)	H(h/d * d/yr)
Acid solution	55	65	12.78	535	2*264
Basic solution	65	80	12.78	802	1*264
Hot sanitary water	10	60	0.78	162	18*264

Table 2.2. *Heat demands in the dairy industry*

Effluents from a concentrator are continuously available at a temperature of 51°C and a mass flow rate of 3.58 kg/s.

To treat this case we need to determine the production schedule, in other terms, the timing of every heat or cooling demand. The cycle of study is one day (24 h). So the variations of demands in time are shown in the following tables.

Period(s)	0–64,800	64,800–86,400
Q (kW)	162	0
CP(kW/K)	3.26	0

Table 2.3. *Variation of heat sanitary water demands in time*

Period (s)	0–1200	1200–2700	2700–3900	3900–12600	12600–13800	13800–40500	40500–41700	41700–43200	43200–44400	44400–49500	49500–50700	50700–86400
Q(kW)	802	0	802	0	802	0	802	0	802	0	802	0
CP(kW/K)	53.4	0	53.4	0	53.4	0	53.4	0	53.4	0	53.4	0

Table 2.4. *Variation of basic solution heating demands in time*

Period (s)	0–600	600–2700	2700–3300	3300–14400	14400–15000	15000–44100	44100–44700	44700–46800	46800–47400	47400–53100	53100–53700	53700–86400
Q(kW)	535	0	535	0	535	0	535	0	535	0	535	0
CP(kW/K)	53.4	0	53.4	0	53.4	0	53.4	0	53.4	0	53.4	0

Table 2.5. *Variation of acid solution heating demands in time*

To treat this case, we used the formulation described in section 2.4. The problem will be solved with the following assumptions:

– the minimum temperature difference of the exchangers between the circulating fluid in the network and the operating streams (acid, base, HSW and effluents) is considered constant at 5 K;

– the efficiency of heat pumps with respect to the Carnot cycle, when used, is 0.5;

– the number of periods is 23 and their durations are shown in Figure 2.19;

– the number of streams that need to be cooled is NFC = 1 (concentrator effluent) and those that need heating are NFF = 3;

– the type of energy conversion systems used is heat pumps;

– the minimum temperature difference at the condenser and evaporator of heat pumps is considered constant at 5K;

– the temperature of the hot source for hot utility is 700°C and the temperature of the cold source for cold utility is 5°C, and the reference temperature for exergy calculations is 5°C;

– the fluid used to store heat is water, its Cp is considered constant at 4,180 kJ/K.kg;

– the number of working days is 264 days.

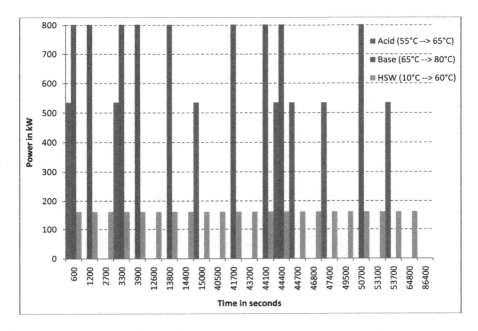

Figure 2.19. *Variation of heat demands in the dairy industry in time.*
For a color version of this figure, see www.iste.co.uk/zoughaib/pinch.zip

2.14.1. *Studied scenarios*

Different scenarios are studied. First, a scenario where the number of heat storage tanks and the number of heat pumps are not limited, which allows us to determine an ideal solution setting the exergy consumption at the lowest reference.

The other scenarios study the impact of varying the number of heat storage tanks and heat pumps.

The first optimization results are presented in Figure 2.20. In this figure, the red curve represents the exergy consumption in the integrated process when no limitation on the number of heat storage tanks and heat pumps is imposed. This result is achievable with 13 heat storage tanks and nine heat pumps.

The other results are obtained without limiting the number of heat storage tanks but limiting the number of heat pumps from 0 to 12.

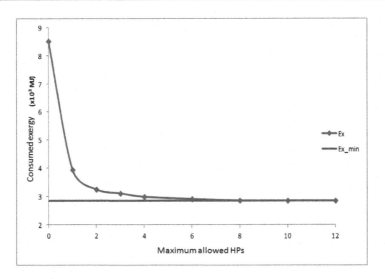

Figure 2.20. *Variation of consumed exergy as a function of allowed heat pumps number. For a color version of this figure, see www.iste.co.uk/zoughaib/pinch.zip*

When no heat pump is allowed, the exergy consumption is high due to the fact that only a limited heat recovery is possible by preheating water using the heat of the concentrator effluent; in that case, no heat storage tank is required since the hot water is produced during a high number of periods. This heat recovery allows 34.6% exergy consumption reduction compared to the situation where no heat integration is done (13×10^3 MJ).

For a number of authorized HPs higher than 6, the value of exergy consumption is equal to the lowest exergy consumption limit.

Table 2.6 presents the results for the 0–3 authorized heat pump scenarios since for more than three HPs the exergy consumption reduction is no longer significant.

Number of heat pumps	Number of heat storage tanks	Consumed exergy ($\times 10^3$ MJ)
0	0	8.52
1	2	3.94
2	4	3.25
3	3	3.11

Table 2.6. *Results of optimization scenarios*

For the case with three HP's and three storage tanks, the architecture proposed by the model is shown in Figure 2.21.

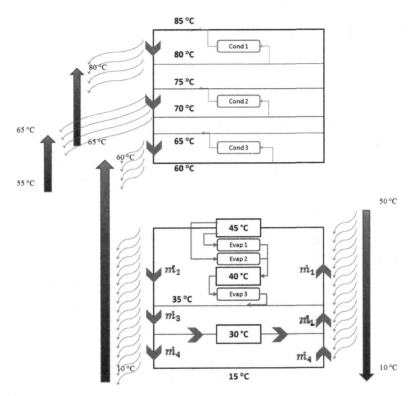

Figure 2.21. *Architecture of the integrated process with three HPs and three storage tanks. For a color version of this figure, see www.iste.co.uk/zoughaib/pinch.zip*

The solution presented in Figure 2.21 suggests three heat storage tanks of the concentrator effluents while the heat pumps convert that low grade heat to the needed temperature for each cold stream. Such architecture efficient in exergy terms will definitely be non-viable from an economic point of view since it leads to investing in high power heat pumps.

Constraining the installed power of heat pumps or introducing an economic objective function instead of the exergy one leads to an equivalent solution in terms of exergy consumption while being less expensive in terms of investment. This solution is presented in Figure 2.22.

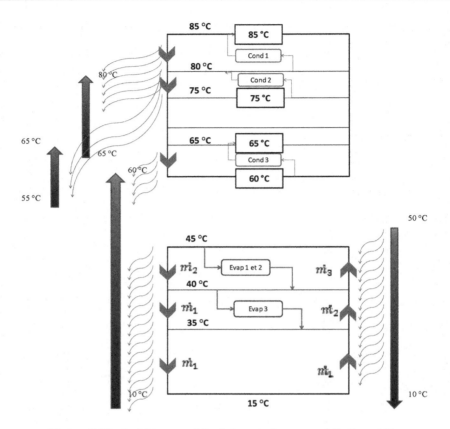

Figure 2.22. *Architecture of the integrated process with three HPs (constraining the power of HPs). For a color version of this figure, see www.iste.co.uk/zoughaib/pinch.zip*

With this solution, four heat storage tanks are needed. These heat storage tanks are placed with high temperature levels suggesting that the concentrator effluents will be used on a continuous basis (maximizing the running time of the heat pumps) thus reducing their power and therefore their investment cost. This equivalent solution is more realistic in practice.

Exergy-based Methodology for Cycle Architecture and Working Fluid Selection: Application to Heat Pumps

The introduction of the exergy concept together with the pinch methodology allowed for both continuous and variable processes (see Chapters 1 and 2) to achieve higher energy use efficiency, thanks to the integration, in addition to the direct heat recovery, of heat conversion systems. Therefore, we can target both the minimum energy requirement and the minimum exergy requirement (MER and MExR).

The methodology presented in both Chapters 1 and 2 uses reduced order models of heat conversion systems based on an exergy efficiency (second law efficiency) and ideal cycle COP and efficiencies. This representation instead of considering market proposed heat conversion systems is done on purpose since:

– when integrating these systems in a process, usually a tailor cut system is best, since considering market systems only limits the possibilities;

– calculation speed is improved (this is a secondary aim but it is helpful if many iterations are needed).

Therefore, integration studies using the methodology described earlier lead to an energy and exergy target that is feasible only if the proposed systems are feasible in terms of operating conditions and their second law efficiency reaches the one assumed for the study.

Trying to match market systems with the specifications coming from these studies may lead to solutions in some cases, but in most of the cases, it is important to achieve an optimized design of these systems.

Here, an exergy-driven methodology also allows for:

– finding the best thermodynamic cycle for the application;

– designing the best working fluid.

To be practical, the methodology formulation will be based on heat pumps; however, the same approach can be used for ORCs and other conversion systems. The reader may consult [AYA 14] for ORCs.

3.1. Methodology description

Many technical options allow the optimization of mechanical vapor compression heat pump systems for the refrigeration process and industrial heating.

The first and most important step is the selection of the most appropriate cycle. This step is driven by exergy destruction limitation mainly in heat exchangers (condenser and evaporator). Once the most promising thermodynamic architecture is preselected, the shortlist of working fluids has to be preselected. This step is a multi-criteria step driven by technical, safety and environmental criteria. Finally, for each working fluid, the operating parameters are pressure, subcooling, internal heat exchanger, refrigerant mixture composition, etc. In this final step, a mathematical model of the heat pump is used for maximizing the COP while defining the best operating parameters.

Therefore, the methodology is a sequential methodology of three steps:

– heat source and sink analysis leading to the preselection of the most interesting cycle architectures;

– preselection of the working fluid shortlist on a multi-criteria basis;

– mathematical optimization of cycle architectures and operating parameters.

In addition to the COP to be maximized, other criteria may be important to analyze:

– technical criteria (pressure levels, pressure ratio);

– safety;

– environment.

3.2. Cycle architecture analysis

Figure 3.1 shows the evolution of temperatures versus enthalpy difference in both heat exchangers of a heat pump for subcritical and transcritical cycles and for three industrial process situations. The nature of the working fluid is also taken into account (pure refrigerant, azeotropic and non-azeotropic mixture). The exergy analysis method is particularly adapted to guide the selection of cycle architectures adapted to process sources and sinks. In a first approach, the irreversibilities on heat exchangers can be assessed by the area between the composite curves. Thus, it can be easy to graphically determine the best cycle for every process situation. Situation (a) is a process where the heating needs and the cooling needs take place at an almost constant temperature (e.g. condensation, evaporation or high mass flow rate). In situation (b), the heating and cooling needs are typical of liquid or gas cooling and heating. Situation (c) is a combination of situations (a) and (b) where the cooling needs take place at an almost constant temperature.

Each of the analyzed cycles has its particularities. The subcritical vapor compression heat pump consists of four mains parts: evaporator (evaporation 1–2' and superheating 2'–2), compressor (2–3), condenser (desuperheating 3–4', condensation 4'–4" and subcooling 4"–4) and an expansion valve (4–1) at subcritical pressures. The superheating ΔT_{SH} (2'–2) is required to protect the compressor by making sure the refrigerant is fully evaporated at the compressor inlet. The ΔT_{SH} parameter can be controlled by the expansion valve. Generally, it is fixed between 3 and 8°C. The subcooling ΔT_{SC} (4"–4) is also a parameter that can be controlled by adjusting the active charge of the heat pump. The refrigerant choice changes the heat pump behavior. For pure refrigerants, the subcritical vapor compression heat pump has no temperature glide on the two HEXs. These heat pumps are adapted for processes of low temperature lift, since the exergy losses at the HEXs are minimized. On the other hand, as shown in Figure 3.1, the use of non-azeotropic mixtures causes temperature glide during the condensation and evaporation which reduces the exergy destruction when the glide of the heat source and heat sink matches the refrigerant one. A transcritical vapor compression heat pump consists of an evaporator (evaporation at subcritical pressure 1–2' and super-heating 2'–2), compressor (2–3), gas cooler (gas cooling at supercritical pressure 3–4) and expansion valve (4–1). Here, the subcooling does not exist because the saturation line of the refrigerant is under the gas cooling curve but the pseudo-subcooling can be used, which is defined as the temperature difference between the pseudo-critical temperature at fixed pressure and the outlet of the gas cooler. The definition of the pseudo-critical temperature T_{pc} is the temperature at which the specific heat reaches a maximum for a given pressure. The active charge defines the high pressure and the pseudo-subcooling of the cycle, which has a significant impact on the performance of a transcritical vapor compression heat pump. As shown in Figure 3.1, when the refrigerant is a pure fluid or an azeotropic mixture, the transcritical heat pumps present less exergy losses for high-glide heat sinks and small-glide heat sources. For high-glide heat sinks and heat sources, a non-azeotropic mixture may be expected to better match the different glides.

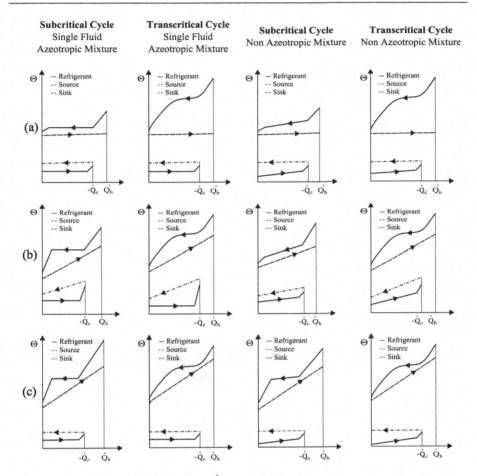

Figure 3.1. *Diagram* $(\Theta;\dot{Q})$ *showing the HEXs of four vapor compression heat pump cycles for three application processes (a, b and c); the HEXs are considered in counter flow*

3.3. Working fluid selection criteria

The number of applicable working fluids gives a large number of freedom degrees to the thermodynamic cycle of heat pump design. Before using the mathematical model for steady-state simulation based on the pinch method (third step of the methodology), a shortlist of candidate working fluids to be used in the application (alone or in a mixture) has to be created from a large number of existing molecules (Table 3.1).

Refrigerants	Critical pressure (MPa)	Critical temperature (°C)	GWP[1] 100-year	Type	Safety group	Nature
Methane	4.6	−82.59	–	Natural	A3	Dry
CO_2	7.38	30.98	1	Natural	A1	Dry
Ethane	4.87	32.18	–	Natural	A3	Dry
Propane	4.25	96.65	3.3	Natural	A3	Isentropic
Butane	3.8	152.05	4	Natural	A3	Wet
Water	22.06	373.95	<1	Natural	A1	Dry
R41	5.9	44.13	150	HFC	A1	Dry
R125	3.63	66.18	3,500	HFC	–	Isentropic
R143a	3.76	72.71	3,800	HFC	A2L	Isentropic
R32	5.78	78.11	670	HFC	A2L	Dry
R134a	4.06	101.06	1,300	HFC	A1	Isentropic
R227ea	2.92	101.75	2,900	HFC	A1	Wet
R152a	4.52	113.26	132	HFC	A2	Dry
R236fa	3.2	125.55	6,300	HFC	–	Wet
R236ea	3.41	139.22	–	HFC	–	Wet
R245fa	3.64	154.05	820	HFC	B1	Wet
R338mccq	2.73	158.8	–	HFC	–	Wet
R245ca	3.92	174.42	560	HFC	–	Wet
RE245cb	2.89	133.68	–	HFE	–	Wet
RE347mcc	2.48	164.55	-	HFE	–	Wet
R1234yf	3.38	94.7	4	HFO	A2L	Wet
R1234ze-(E)	3.63	109.36	–	HFO	A1	Wet
R218	2.68	71.89	7,000	PFC	A1	Wet
RC318	2.78	115.23	8,700	PFC	A1	Wet

Table 3.1. *Available refrigerants (non-exhaustive)*

To determine the shortlist, the first criteria is the thermodynamic properties that should be in accordance with the heat source and sink temperature levels and the preselected cycle architectures (e.g. if the selected

---

1 GWP: Global warming potential.

cycle architecture is a transcritical one, the critical temperature of the fluid should not be too high compared to the heat sink). In addition to the information present in Table 3.1, we should be interested in the saturation pressure relation with temperature.

The methodology of fluid design is a multi-objective and multi-constraint optimization process. Therefore, in addition to the thermodynamic properties that are the main criteria, we should consider:

– *Environmental criteria:* The working fluid must have a low environmental impact represented by the GWP. This aspect is controlled by standards in several countries (e.g. F-gaz in the EU).

– *Safety criteria*: It concerns either the toxicity (B safety group) or the flammability of the refrigerant. To distinguish a flammable fluid from a non-flammable one, a coefficient called *RF-number*, defined by Kondo [KON 02], is used.

$$RF = \frac{(LFL * UFL)^{1/2} - LFL}{LFL} * \frac{P_c}{P_m}, \qquad [3.1]$$

where *RF* is the RF number; *UFL* is the upper flammability level; *LFL* is the lower flammability level; P_c is the heat of combustion of the fluid; M_m is the molar mass of the fluid.

For this evaluation, fluids are considered moderately flammable (A2L) if their RF number is less than 30 and highly flammable (A2, A3) if their RF number is greater than 30.

– *Compressor discharge temperature*: This temperature should be limited in order to use the actual lubricants. Most of the actual lubricants withstand temperatures up to 130°C while we may find the lubricants adapted to higher temperature levels. It is directly related to the γ factor ($\gamma = C_p/C_v$). The shape of the saturation curve helps identify three groups of fluids (Figure 3.2).

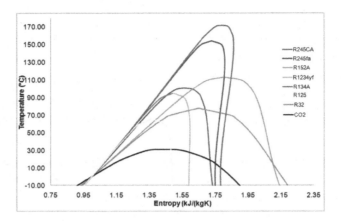

Figure 3.2. *Saturation curves of some fluids. For a color version of this figure, see www.iste.co.uk/zoughaib/pinch.zip*

When the vapor saturation curve has a negative slope (e.g. in the case of CO_2, R32), the refrigerant γ factor and the compression discharge temperature are relatively high. This group of fluid is called dry since the compressor discharge will be in the superheated vapor state when the compressor inlet is in the saturated vapor state. Some refrigerants have an almost vertical vapor saturation curve (R134a, R152a); these refrigerants have a smaller γ factor and the temperature increase during compression is limited. They are called isentropic fluids. The last category presents a positive slope vapor saturation curve (e.g. R245fa, R1234yf). These refrigerants present a very low γ close to 1. If the compression starts from a vapor saturated state, the compressor discharge state is in the two-phase region. These fluids are called wet and need high superheat at the compressor inlet to avoid two-phase discharge.

The properties of fluids also help identify the impact of the introduction of internal heat exchanger (a heat exchanger that increases the subcooling while superheating the vapor at the outlet of the evaporator). It is obvious to see that the internal heat exchanger (IHX) should be avoided for dry fluids.

– *Cycle high pressure*: This pressure should be limited to avoid high tubing costs. High pressure levels (> 30 bars) require non-standard elements.

– *Swept volume of the compressor*: Minimizing the swept volume of the compressor reduces the investment costs. The pressure level at the heat source temperature is a good indicator on the compressor swept volume, since most of the compressors used for heat pumps are volumetric one (except the centrifugal compressor family). Hence, the higher the pressure, the higher the density and the smaller the swept volume for a certain refrigerant mass flow rate is. However, too high a pressure at the inlet of the compressor means high discharge pressure. Therefore, a compromise between these criteria may be needed. Figure 3.3 shows the saturation pressure as a function of the saturation temperature of some fluids.

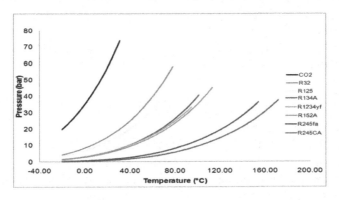

Figure 3.3. *Saturation pressure of some fluids. For a color version of this figure, see www.iste.co.uk/zoughaib/pinch.zip*

– *Mixture of fluids*: As shown in section 3.1.1, in a certain heat sink and source configuration, it may be interesting from a thermodynamic point of view to blend fluids creating a non-isentropic mixture with a certain controlled temperature glide. Creating a mixture of fluid can also help fit some other criteria such as environmental one or safety one.

When preselecting the shortlist of fluids, a fast method allows *a priori* to find a good mixing candidate. Indeed, many criteria or properties are determined as a linear combination of the mass fraction (GWP, RF) and others can be approximated by a linear combination. The temperature glide, for instance, can be related to the difference in critical temperature of the components and the composition, and the saturation pressure is a function of each fluid saturation pressure at the same temperature and of the composition. When more than two fluids are mixed (usually 3), a graphical method (a triangle graph being used for three components) allows finding the composition of the mixture for some iso properties. This will be explained more in detail in the case study presented in section 3.2.

3.4. Mathematical optimization of cycle architecture and operating parameters

Once the adapted cycle architectures and a shortlist of fluids are selected, the third step is to determine accurately for each cycle and working fluid candidate the best performance values for the evaluation criteria while optimizing the operating conditions (e.g. pressure level, composition, use of IHX, subcooling). This is done using a mathematical model of the heat pump.

The approach used to model the heat pump behavior in a steady state is commonly called the "pinch" method (different from the pinch analysis method but based on similar criteria). It consists of defining the minimum temperature difference between the refrigerant and the medium in the heat exchangers. With energy balances (first law), and the "pinch" conditions, the high and low pressure may be determined. The cycle can be determined and the refrigerant mass flow rate can be calculated. Depending on the flow arrangement of the heat exchangers (cross, parallel or counter flow), the thermodynamic region of the pinch point exists at different locations. For the parallel flow heat exchangers, the pinch point is always located at the outlet. For the cross flow heat exchangers, the pinch point is located between the inlet cold fluid and the outlet hot fluid. However, the counter flow heat exchanger has two possible pinch points: one at the inlet cold fluid and another at a location that depends on the heat exchanger function (evaporator, condenser or gas cooler). The different possible locations of the two pinch points for counter flow heat exchangers are presented in Figure 3.4. For the evaporator, the second possible pinch point is located at the outlet of the cold source. For the condenser, the possible second pinch point is located at the refrigerant vapor saturation line. For the gas cooler, the second pinch point may be located everywhere between the refrigerant inlet and the pseudo-critical point.

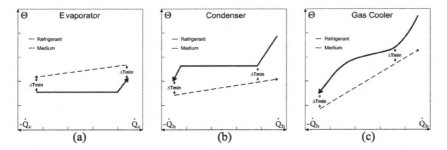

Figure 3.4. *Possible pinch point locations for the counter flow heat exchangers*

3.4.1. *Fluid property implementation*

The refrigerant properties as well as the process fluids (the heat sink and source) should be calculated using a thermodynamic model. Many programs of process simulation software include a library of thermodynamic models for the refrigerant properties, but we can also use dedicated libraries such as Refprop (from the NIST or Coolprop edited by the University of Liège).

The functions needed for building the cycle are:

– P_{sat} function of T;

– T_{sat} function of P;

– s function of P, h;

– s function of P, T;

– h function of P, T;

– h function of P, Q;

– T function of P, h;

– T function of P, Q;

– ρ function of P, h.

3.4.2. *Assumptions*

For a pre-design optimization means, the following assumptions can be made:

– each component of the cycle is analyzed as a control volume at the steady state;

– there is no heat exchange between each component with its surroundings;

– there are no pressure drops through the heat exchangers;

– the heat exchangers are counter flow ones (in most industrial situations);

– the compressor (volumetric technology) operates adiabatically with constants isentropic and volumetric efficiency;

– the expansion through the valve is a throttling process;

– kinetic and potential energy are negligible.

3.4.3. *Performance model*

The energy efficiency of the heat pump is represented by the heating COP, which is defined as follows:

$$COP = \frac{\dot{Q}_{HEX}^{HP}}{\dot{W}_{in}} \qquad [3.2]$$

where $\dot{Q}_{HEX}^{HP}$ is the heat transfer rate at the gas cooler or condenser and $\dot{W}_{in}$ the input power of the compressor. In order to determine the exergy efficiency of the heat pump (second law efficiency), the ideal COP is calculated. Because of the variation of the source and sink temperatures, the entropy-based average temperature is defined as the ratio between the enthalpy variation and the entropy variation:

$$\tilde{T} = \frac{h_{out} - h_{in}}{s_{out} - s_{in}} \qquad [3.3]$$

where $\tilde{T}$ is the average temperature in the sense of the entropy variation, h is the specific enthalpy and s is the specific entropy. The exergy efficiency of the heat pump can be defined as:

$$\eta_{HP} = 1 - \frac{\sum \zeta}{\dot{W}_{in}}$$ [3.4]

where ζ is the exergy destruction in the HP components. When $T_0 = \tilde{T}_{c_{me}}$ (T_0 is the reference temperature for exergy calculation and $\tilde{T}_{c_{me}}$ is the average temperature in the sense of entropy variation of the cold source), the exergy efficiency η_{HP} is equal to the second law efficiency and can be defined as:

$$\eta_{HP}\left(T_0 = \tilde{T}_{cme}\right) = \left(1 - \frac{\tilde{T}_{c_{me}}}{\tilde{T}_{h_{me}}}\right) * \frac{\dot{Q}_{HEX}^{HP}}{\dot{W}_{in}}$$ [3.5]

where T_0 represents the reference temperature of the system, $\tilde{T}_{c_{me}}$ is the entropy-based average temperature of the process heat source and $\tilde{T}_{h_{me}}$ is the entropy-based average temperature of the process heat sink.

3.4.4. The subcooling parameter

The following theoretical study shows, on the one hand, how subcooling increases the coefficient of heat pump performance. And, on the other hand, how to easily determine the optimal value of the subcooling temperature according to the operating conditions.

Enthalpy variation of a subcritical condenser is shown in Figure 3.5 to facilitate the comprehension of the demonstration. However, this demonstration is also valid for a supercritical gas cooler. The terms $T1_m$ and $T4_m$ shown in Figure 3.5(a) and (b) are input parameters, defining the operating conditions of the process, namely the input and output temperatures of the heated medium. The parameter ΔT_3 shown in Figure 3.5(a) and (b) defines the minimum temperature difference between the medium and the refrigerant in the condenser or gas cooler, frequently called the pinch point. Its physical position depends on subcooling (pseudo-critical subcooling for the gas cooler) and the medium temperature. The following assertions are used for the demonstration:

– the compressor has a constant isentropic efficiency. Thus, if the outlet temperature of the refrigerant at the evaporator is fixed, the isentropic curve is fixed, as shown in Figure 3.5(a);

– to simplify, at the evaporator, the medium is considered as a heat source with an almost constant temperature.

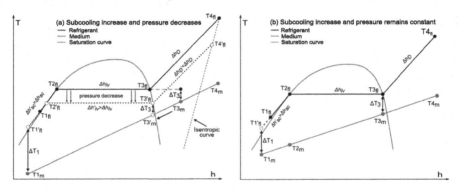

Figure 3.5. *Evolution of the composite curves when the subcooling increases. For a color version of this figure, see www.iste.co.uk/zoughaib/pinch.zip*

The first step of the demonstration is to show that when the subcooling increases, the pressure of the refrigerant in the condenser or in the gas cooler also increases. Indeed, when subcooling increases, there are three possibilities that the pressure varies. Either the pressure of the fluid decreases or remains constant or increases.

The pressure would decrease when subcooling increases, as shown in Figure 3.5(a) where the dotted line represents the new isobaric curve and the prime (′) denotes the new temperatures. Thus, the saturation temperature and $T3_m$ decrease due to the pinch condition. Thus, Δh_{lv} increases, because for all fluids, the latent heat is a decreasing function of the temperature. So the sum $\Delta h_{lv} + \Delta h_{sc}$ increases. Therefore, to keep the heat power balance with the medium between point 1 and point 3, it is necessary that the mass flow rate of the refrigerant decreases to compensate the increase of the enthalpy variation. Moreover, at each reduction of the saturation temperature, the enthalpy variation Δh_D decreases because the derivative of the isentropic curve with respect to temperature is always larger than the derivative of the saturation vapor curve with respect to temperature. Hence, $T3_m$ decreases and $T4_m$ and $\dot{m}_m$ are fixed by the operating conditions. Thus, the heat exchanged between points 3 and 4 increases, which is in contradiction with the reduction of the refrigerant mass flow rate and the diminution of Δh_D. Therefore, the pressure cannot decrease when ΔT_1 increases.

Assume now that the pressure remains constant when subcooling increases, as shown in Figure 3.5(b). Therefore, Δh_{lv} and Δh_D remain constant because the saturation temperature and $T4_{fl}$ are fixed. Thus, the mass flow rate of the refrigerant must decrease because the enthalpy variation Δh_{sc} is added to the power balance, which is in contradiction with the heat power balance between points 3 and 4.

Therefore, the pressure can only increase when subcooling increases.

When the subcooling is non-existent, the pinch position is at the saturation vapor line for a condenser or above the pseudo-critical temperature at the change of slope for a gas cooler. As shown in Figure 3.6, during the increase of subcooling, at a certain subcooling level, the pinch changes its position to the outlet of the condenser.

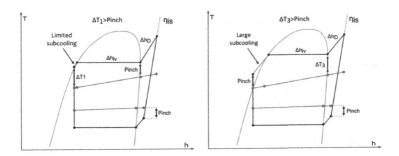

Figure 3.6. *Heat pump cycle in a (T, h) diagram with representation of the possible pinch positions on the condenser and the evaporator. For a color version of this figure, see www.iste.co.uk/zoughaib/pinch.zip*

According to the position of the pinch, the increase of subcooling has different impacts on the pressure. When the pinch is placed at the vapor saturation point (Figure 3.6(a)), the saturation temperature is locked by the pinch condition and the medium temperature. At each rise of subcooling, the pressure increases but very slightly. On the other hand, when the pinch is placed at the outlet of the condenser (Figure 3.6(b)), the rise of subcooling implies an identical rise of the saturation temperature and pressure.

In accordance with Figure 3.7, the COP could be written as in equation [3.6]:

$$COP = \frac{\dot{Q}_h}{\dot{Q}_h - \dot{Q}_{evap}} = \frac{\varepsilon_{sc} + \Delta h_i + \varepsilon_D}{\Delta h_i - \Delta h_e + \varepsilon_D} \qquad [3.6]$$

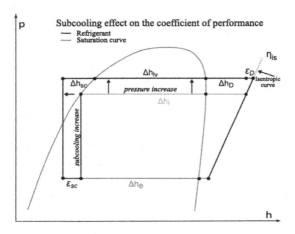

Figure 3.7. *Subcooling effect on a heat pump in a Mollier diagram.*
For a color version of this figure, see www.iste.co.uk/zoughaib/pinch.zip

While the pinch is at the vapor saturation position, ε_{sc} increases with the subcooling while ε_D increases very slightly. Therefore, the COP is a function mainly of ε_{sc} and the derivative with respect to ε_{sc} [3.7] is always positive:

$$\frac{dCOP}{d\varepsilon_{sc}} = \frac{1}{\Delta h_i - \Delta h_e + \varepsilon_D} \geq 0 \qquad [3.7]$$

When subcooling becomes large enough to imply a new position of the pinch, as shown in Figure 3.6(b), each increase of subcooling implies an increase of the saturation pressure, so only ε_D increases. Therefore, the COP is a function of only ε_D and the derivative with respect to ε_D [3.8] is always negative:

$$\frac{dCOP}{d\varepsilon_D} = \frac{-(\Delta h_e + \varepsilon_{sc})}{(\Delta h_i - \Delta h_e + \varepsilon_D)^2} \leq 0 \qquad [3.8]$$

In conclusion, the COP rises until the double pinch condition is respected, and this configuration gives the optimal subcooling value. Thus, the double "pinch" condition can be implemented explicitly to optimize the cycle integration.

3.5. Heat pump in a dairy process example

The methodology presented in section 3.1 is applied to the design of one of the heat pumps proposed by the exergy-based multi-period algorithm (see section 2.5.2). This heat pump is the one dedicated to hot water production (HP3). The integration result has shown that before being heated by the heat pump, water is preheated by exchanging heat with the concentrator effluent (through the integration intermediate network). The simplified layout of this subsystem is shown in Figure 3.8.

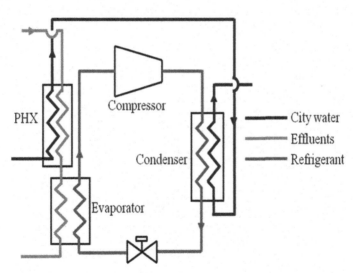

Figure 3.8. *Simplified layout of the integrated subsystem for hot water generation. For a color version of this figure, see www.iste.co.uk/zoughaib/pinch.zip*

From the integration results (Figure 2.22), we can retrieve the temperature evolution of the heated water (process stream) and the water of the indirect heat integration network on the boundary of both the direct heat exchanger (PHX) and the heat pump. These values are given in Table 3.2.

	PHX		Heat pump	
	Inlet T (°C)	Outlet T (°C)	Inlet T (°C)	Outlet T (°C)
Heat integration network	51	30	30	24
Hot water	10	46	46	60

Table 3.2. *Inlet and outlet temperatures*

Table 3.2 shows that the temperature glide of the heat sink and source at the evaporator and the condenser is important; especially on the condenser side. From Figure 3.1, we are in the process situation (b). That is why it may be of interest to use blends of refrigerants with glide matching (non-azeotropic mixtures) to reduce the exergy losses in the heat exchangers which permit to improve the performance of the system.

The pure fluids that may be used as components of a refrigerant blend or solely are first preselected in terms of thermodynamic properties, as presented in Table 3.3.

Symbol	Mm (g/mol)	T_{crit} (°C)	P_{crit} (MPa)	NBT[2] (°C)	GWP (kgeq CO_2)
R-134a	102.03	101.06	4.06	− 26.1	1300
R-152a	66.05	113.26	4.52	− 24	132
R-1234yf	114.04	94.8	3.26	− 29.2	4
R-245fa	134.05	154.1	4.43	15.1	820
R-32	52.02	78.1	3.64	− 51.7	670
R-125	120.02	66.02	3.45	− 48.09	3500
R-744	44.01	30.97	7.38	− 78.4	1

Table 3.3. *Properties of the selected fluids*

From Figures 3.2 and 3.3, we note that R-32 and R-125 have similar properties with a big difference of GWP; then, R-32 will be used as a component of the blend.

---

2 NBT: Normal boiling temperature. It is the saturation temperature at the pressure of 1 bar.

R-1234yf, R-134a and R-152a also have similar properties; then, we can use one of them with other pure fluids. For a certain saturation temperature, R-32 has the highest pressure while R-245fa has the lowest. R-1234yf, R-134a and R-152a present an intermediate pressure. Mixing R-32 with R-245fa allows high temperature glide and intermediate pressure levels. Adding one of R-1234yf, R-134a and R-152a to this mixture allows controlling other aspects such as GWP or flammability.

Using these refrigerants, several blends are compared in terms of condensing pressure, GWP, COP, compressor discharge temperature, swept volume of the compressor and exergy destruction.

The method of selection of mixtures from the pure fluids is based on the triangle method. Figure 3.9(a), (b) and (c) shows an illustration of this representation for three possible ternary mixtures.

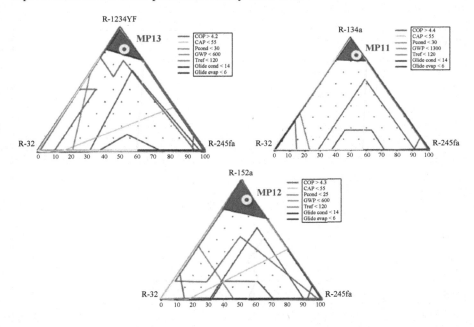

Figure 3.9. *Triangle representation of mixture property zones. For a color version of this figure, see www.iste.co.uk/zoughaib/pinch.zip*

In Figure 3.9, we can see that zones with several constraints are respected (e.g. within the green zone, the high pressure is lower than the fixed limit). The dark gray areas represent the admissible areas where the mixtures respond to all criteria.

The thermodynamic cycle properties and the characteristics of the heat pumps are calculated by considering a constant pinch at the heat exchangers of 5 K.

We should note that the fluid R-152a has higher flammability than R-1234yf; then if two blends have the same composition of their components, and one contains R-152a and the other contains R-1234yf, it is obvious to say that the blend containing R-152a is more flammable than the other. Table 3.4 shows a comparison for the different criteria between three blends (best composition in terms of COP within the gray area for each blend) and four pure refrigerants that can be used for the heat pump.

Refrigerant	MP11	MP12	MP13	R-134a	R-152a	R-1234yf	R-245fa
Mass composition (%)	85% R-134a, 10% R-245fa, 5% R-32	85% R-152a, 10% R-245fa, 5% R-32	85% R-1234yf, 10% R-245fa, 5% R-32	100	100	100	100
P_{cond} (MPa)	1.7	1.55	1.79	1.8	1.6	1.73	0.52
$COP_{heat\ pump}$	4.24	4.26	4.27	3.8	4	3.6	3.97
SV (m³/h)	52	58	51	56.4	58.4	63.2	196
$T_{outcomp}$ (°C)	83	92	77	83	93	75	76
GWP	1220	228	119	1300	132	4	820
Ex destruction (kW)	7.63	7.42	7.27	8.86	8.44	9.3	8.13
Evap glide (k)	4.5	3.43	7.28	0	0	0	0
Cond glide (k)	4.37	3.23	5.26	0	0	0	0

Table 3.4. *Results of selected blends and pure fluids for the heat pump*

The results of Table 3.4 confirm that the preselected cycle configuration (subcritical with non-azeotropic mixture) is the one permitting the highest COP compared to the traditional pure refrigerant subcritical cycle. For the exergy destruction and the COP, very close results are achieved by the three proposed mixtures (for the best composition in terms of COP).

However, while looking at the other criteria, the highest difference is:

– GWP: where MP11 achieves the highest GWP while MP13 has the lowest one. For the pure refrigerants, the best result is for R-1234yf but with 15% lower COP;

– flammability: MP11 is the least flammable while MP13 presents the intermediate flammability;

– compressor swept volume: MP13 needs the smallest compressor.

This multi-criteria analysis ultimately favors the MP13 mixture.

Conclusion

This book reported some recent developments of process integration and design based on the historical pinch method. The main novelty comes from the coupling of the pinch method with the exergy analysis, leading to new possibilities and further energy efficiency increase.

Heat storage integration into variable and batch processes allows greater heat recovery and adds flexibility to the processes.

The exergy concept can hence be used at several scales:

– analyzing the process operations and selecting operating parameters that may be optimized and technical options to be considered;

– when performing the pinch analysis or the multi-period pinch analysis considering in a systematic way the energy conversion systems (and heat storage for variable processes case);

– optimizing the detailed design of the energy conversion systems by selecting the cycle architecture and the working fluid.

This methodology is fully implemented in the open-source platform CERES developed by the CES of Mines ParisTech.

Currently, the methodology is further developed trying to work on the scale of large industrial sites and industrial territories. These works extend the work presented in this book by introducing the energy transport systems needed at this scale.

When many industrials are involved in an eco-industrial park approach, economic governance modeling and robustness assessment become necessary otherwise it can be difficult to convince decision-makers to approve such synergy solutions.

All these developments will be the subject of a future book.

Bibliography

[ANA 10] ANANTHARAMAN R., NASTAD I., NYGREEN B. *et al.*, "The sequential framework for heat exchanger network synthesis – the minimum number of units sub-problem", *Computers and Chemical Engineering*, vol. 34, no. 11, pp. 1822–1830, 2010.

[AYA 14] AYACHI F., BOU LAWZ KSAYER E., ZOUGHAIB A. *et al.*, "ORC optimization for medium grade heat recovery", *Energy*, vol. 68, pp. 47–56, 2014.

[BAR 05] BARBARO A., BAGAJEWICZ M., "New rigorous one-step MILP formulation for heat exchanger network synthesis", *Computers and Chemical Engineering*, vol. 29, no. 9, pp. 1945–1976, 2005.

[BEC 12a] BECKER H., Methodology and thermo-economic optimization for integration of industrial heat pumps, PhD Thesis, EPFL, 2012.

[BEC 12b] BECKER H., MARÉCHAL F., "Energy integration of industrial sites with heat exchange restrictions", *Computers and Chemical Engineering*, vol. 37, pp. 104–118, 2012.

[BIE 03] BIELER P.S., FISCHER U., HUNGERBUHLER K., "Modeling the energy consumption of chemical batch plants: bottom-up approach", *Industrial and Engineering Chemistry Research*, vol. 42, pp. 6135–6144, 2003.

[CGE 14] CGE/CGEDD/IGF, Les certificats d'économies d'énergie: efficacité énergétique, Report, French Ministry for Economy and Finance, 2014.

[CLA 86a] CLAYTON R.W., "Cost reductions on a productions of synthetic resins by a process integration study at Cray Valley Products Ltd", Report, Energy Technology Support Unit (ETSU), 1986.

[CLA 86b] CLAYTON R.W., "Cost reductions on an edible oil refinery identified by a process integration study at Van den Berghs and Jurgens Ltd", Report, Energy Technology Support Unit (ETSU), 1986.

[DIN 02] DINCER I., ROSEN M.A., *Thermal Energy Storage: Systems and Applications*, John Wiley & Sons, Chichester, 2002.

[DOE 16] DOE, International energy outlook 2016, Report, U.S. Energy Information, 2016.

[FUR 01] FURMAN K., SAHINIDIS N., "Computational complexity of heat exchanger network synthesis", *Computers and Chemical Engineering*, vol. 25, pp. 1371–1390, 2001.

[FUR 02] FURMAN K., SAHINIDIS N., "A critical review and annotated bibliography for heat exchanger network synthesis in the 20th century", *Ind. Eng. Chem. Res.*, vol. 41, pp. 2335–2370, 2002.

[IEA 15a] IEA, World energy outlook special report: energy climate and change, Report, Directorate of Global Energy Economics, 2015.

[IEA 15b] IEA, Energy technology perspectives 2015, Report, IEA/OECD, 2015.

[IEA 15c] IEA, Energy efficiency market report, Report, Directorate of Sustainable Energy Policy and Technology, 2015.

[IEA 16] IEA, Key world energy statistics, Report, International Energy Agency, 2016.

[JEZ 03] JEZOWSKI J.M., SHETHNA H.K., CASTILLO F.J.L., "Area target for heat exchanger networks using linear programming", *Industrial & Engineering Chemistry Research*, vol. 48, no. 8, pp. 1723–1730, 2003.

[KEM 07] KEMP I.C., *Pinch Analysis and Process Integration: A User Guide on Process Integration for the Efficient Use of Energy*, Institution of Chemical Engineers, 2007.

[KEM 89a] KEMP I.C., DEAKIN A.W., "The cascade analysis for energy and process integration of batch processes. I: Calculation of energy targets", *Chemical Engineering Research and Design*, vol. 67, pp. 495–509, 1989.

[KEM 89b] KEMP I.C., DEAKIN A.W., "The cascade analysis for energy and process integration of batch processes. II: Network design and process scheduling", *Chemical Engineering Research and Design*, vol. 67, pp. 510–516, 1989.

[KEM 90] KEMP I.C., *Process Integration: Process Change and Batch Processes*, ESDU International, London, 1990.

[KLE 13] KLEMEŠ J., VARBANOV P., KRAVANJA Z., "Recent developments in process integration", *Chemical Engineering Research and Design*, vol. 91, no. 10, pp. 2037–2053, 2013.

[KON 02] KONDO S., TAKAHASHI A., TOKUHASHI K. *et al.*, "RF-number as a new index for assessing combustion hazard of flammable gases", *Journal of Hazardous Materials*, vol. A93, pp. 259–267, 2002.

[LIN 83] LINNHOFF B., HINDMARSH E., "The pinch design method for heat exchanger networks", *Chemical Engineering Science*, vol. 38, no. 5, pp. 745–763, 1983.

[MUR 11] MURR R., THIERIOT H., ZOUGHAIB A. *et al.*, "Multi-objective optimization of a multi water-to-water heat pump system", *Applied Energy*, vol. 88, pp. 3580–3591, 2011.

[OEC 14] OECD, OECD Environmental Outlook to 2050: the consequences of inaction key, Report, OECD and the PBL Netherlands Environmental, 2014.

[PAP 83] PAPOULIAS S.A., GROSSMANN I.E., "A structural optimization approach in process synthesis II: Heat recovery networks", *Computers and Chemical Engineering*, vol. 7, no. 6, pp. 707–721, 1983.

[POU 14] POURANSARI N., MARÉCHAL F., "Heat exchanger network design of large-scale industrial site with layout inspired constraints", *Computers and Chemical Engineering*, vol. 71, pp. 426–445, 2014.

[SAL 15] SALAME S., SAAB J., ZOUGHAIB A. *et al.*, *Optimal Design of a Hybrid Air Conditioning System under Electrical Grid Constraint*, ICR, Yokohama, 2015.

[SAL 16] SALAME S., ZOUGHAIB A., "Heat integration in regeneration process of molecular sieves including heat storages", *Internation Journal of Thermodynamics*, vol. 19, no. 3, pp. 137–144, 2016.

[SHA 97] SHARRAT P.S., *Handbook of Batch Process Design*, Springer, 1997.

[SHE 95] SHENOY U.V., *Heat Exchanger Network Synthesis: Process Optimization by Energy and Resource Analysis*, Gulf Publishing Company, Houston, 1995.

[STO 95] STOLZE S., MIKKELSEN J., LORENTZEN B. *et al.*, "Waste-heat recovery in batch processes using heat storage", *Journal of Energy Resources Technology*, vol. 117, pp. 142–149, 1995.

[THI 13] THIBAULT F., ZOUGHAIB A., JUMEL S., "An exergy-based LP algorithm for heat pump integration in industrial processes", *Proceedings of ECOS2013*, 2013.

[THI 15] THIBAULT F., ZOUGHAIB A., PELLOUX-Prayer S., "A MILP algorithm for utilities pre-design based on the Pinch Analysis and an exergy criterion", *Computers and Chemical Engineering*, vol. 75, pp. 65–73, 2015.

[TRA 15] TRAN C.-T., THIBAULT F., THIERIOT H. *et al.*, *New Features to Barbaro's Heat Exchanger Network Algorithm: Heat Exchanger Technologies and Waste Heat Flow Representation*, ECOS, Pau, 2015.

[UN 15] UN, World Population Prospects: the 2015 revision, Report, Department of Economic and Social Affairs Population Division, 2015.

Index

Printed in the United States
By Bookmasters